The Trappers Handbook

Mastering the Art of Ethical Trapping, Modern Hunting, Bushcraft Skills, Skinning, Tanning, and Drying Strategies.

John C. Kunz

All rights reserved. No part of this book may be reproduced, stored in a retrieval system, or transmitted in any form or by any means—electronic, mechanical, photocopy, recording, or otherwise—without the prior written permission of the publisher, except in the case of brief quotations embodied in critical articles or reviews.

Copyright © 2024 by John C. Kunz

Table of Contents

Introduction 6

Chapter ONE 8

Introduction to Ethical Trapping and Modern Hunting 8

    1.1 The History and Evolution of Trapping and Hunting 8

    1.2 Ethical Considerations and Legal Regulations 13

    1.3 Importance of Wildlife Conservation 19

Chapter two 28

Tools and Equipment 28

    2.1 Essential Trapping Tools and Their Uses 28

    2.2 Modern Hunting Gear and Technology 36

    2.3 Maintenance and Safety of Equipment 44

Chapter three 50

Trapping Techniques and Methods 50

    3.1 Types of Traps and Their Applications 50

    3.2 Setting and Baiting Traps Effectively 57

    3.3 Monitoring and Checking Traps 62

Chapter four 68

Modern Hunting Strategies 68

    4.1 Tracking and Stalking Techniques 68

    4.2 Using Calls and Decoys 74

    4.3 Best Practices for Ethical Hunting 79

Chapter five 85

Handling and Processing Caught Animals 85

5.1 Initial Handling and Field Dressing 85

5.2 Transporting Game Safely..90

5.3 Preparing for Skinning and Butchering 95

Chapter six 101

Skinning Techniques..101

6.1 Skinning Tools and Their Uses.......................................101

6.2 Step-by-Step Skinning Process.......................................107

6.3 Handling Different Types of Fur and Hide 111

Chapter seven 117

Tanning and Preserving Hides 117

7.1 Introduction to Tanning Methods 117

7.2 Natural vs. Chemical Tanning 123

7.3 Step-by-Step Tanning Process..128

Chapter eight 134

Drying and Storing Tanned Hides..134

8.1 Methods of Drying Hides 134

8.2 Preventing Mold and Damage in Hides 140

8.3 Long-Term Storage Solutions for Hides........................145

Chapter nine 151

Advanced Techniques and Projects...151

9.1 Creating Leather from Tanned Hides............................151

9.2 Crafting with Fur and Leather 156

9.3 Ethical Trapping and Hunting Community Involvement
....162

Introduction

The art of trapping and hunting has been a fundamental aspect of human survival and culture for millennia. From the earliest hunter-gatherer societies to modern outdoor enthusiasts, the practice of harvesting wildlife has evolved significantly. Today, ethical trapping and hunting are not only about procuring food and resources but also about maintaining a balance with nature, conserving wildlife populations, and respecting the environment.

In the following chapters, you will find detailed information on the history and evolution of trapping and hunting, the essential tools and technologies involved, and the critical ethical considerations that must guide all such endeavors. We will delve into practical techniques for setting and monitoring traps, tracking and hunting game, and processing animals through skinning, tanning, and drying methods. Additionally, this guide emphasizes community involvement and the importance of fostering a culture of conservation and respect for wildlife.

The journey begins with a look back at the origins and development of trapping and hunting, exploring how these practices have adapted over time to meet changing societal needs and ethical standards. From there, we will discuss the modern tools and technologies that have revolutionized these activities, making them more efficient and humane. Essential to this discussion is an understanding of the ethical considerations and legal regulations that ensure these practices are conducted with respect for animal welfare and ecological balance.

As we move through the practical aspects of trapping and hunting, you will learn about different types of traps and their applications, effective baiting and monitoring techniques, and the best practices for humane and ethical hunting. Detailed instructions on handling and processing game will guide you through the steps of field dressing, skinning, tanning, and drying, ensuring that you can make the most of the resources provided by each animal.

Finally, this guide highlights the importance of community involvement in promoting ethical trapping and hunting practices. By engaging with educational programs, advocacy efforts, and collaborative projects, you can contribute to a culture of responsible wildlife management and conservation.

Whether you are a seasoned hunter looking to refine your skills or a newcomer eager to learn the ropes, this guide aims to be your comprehensive resource for mastering the art of ethical trapping and hunting. Through careful study and practice, you can develop the expertise needed to participate in these time-honored traditions with integrity and respect for the natural world.

CHAPTER ONE

Introduction to Ethical Trapping and Modern Hunting

1.1 The History and Evolution of Trapping and Hunting

Early Beginnings: Prehistoric Era

Trapping and hunting have been integral to human survival since the dawn of our species. Early humans relied on these practices to secure food, clothing, and tools. Archaeological evidence suggests that our ancestors used rudimentary tools made from stone, bone, and wood to hunt and trap animals. They developed a deep understanding of animal behavior and the environment, which allowed them to devise effective hunting strategies and trapping techniques.

Ancient Civilizations: Advancements and Innovations

As human societies evolved, so did their methods of trapping and hunting. Ancient civilizations, such as the Egyptians, Greeks, and Romans, made significant advancements in these practices. They developed more sophisticated tools and weapons, such as bows and arrows, spears, and nets. These innovations increased their efficiency and success rates in capturing game.

In addition to technological advancements, ancient cultures also integrated hunting and trapping into their societal and religious practices. For instance, the Egyptians believed that hunting was a divine activity, often depicted in their art and mythology. Similarly, the Greeks and Romans held grand hunting expeditions, known as venationes, which were both a form of entertainment and a display of skill and bravery.

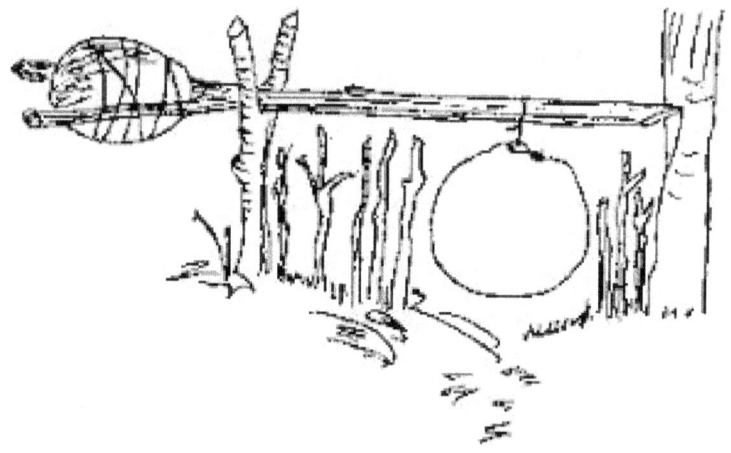

Fig 1: Ancient Trapping Method

Middle Ages: Feudal Systems and Game Management

During the Middle Ages, hunting and trapping became highly regulated activities, particularly in Europe. The feudal system granted the nobility exclusive rights to hunt on vast tracts of land, while commoners were often restricted or entirely prohibited from hunting. This period saw the emergence of game management practices, where the aristocracy implemented measures to ensure sustainable hunting. These measures included the establishment of game reserves, the enforcement of hunting seasons, and the regulation of trap types and methods.

Falconry, the practice of hunting with trained birds of prey, also gained popularity during this era. It was considered a noble sport, requiring significant skill and dedication. Falconry further illustrated the intricate relationship between humans and animals, showcasing the ability to train and collaborate with wildlife for hunting purposes.

Colonial Era: Expansion and Exploitation

The colonial era marked a significant expansion of hunting and trapping activities, driven by European exploration and colonization. Colonists brought their hunting traditions and techniques to the New World, where they encountered diverse and abundant wildlife. Trapping became a lucrative industry, particularly in North America, where fur trading flourished. European settlers and indigenous peoples alike engaged in trapping beavers, otters, and other fur-bearing animals, leading to the establishment of trade networks and economic growth.

However, this period also witnessed the over-exploitation of wildlife populations. The demand for fur and other animal products led to the decline and, in some cases, the extinction of certain species. The passenger pigeon, once numbering in the billions, was driven to extinction due to unregulated hunting and habitat destruction.

19th and 20th Centuries: Conservation and Regulation

The 19th and 20th centuries brought significant changes to trapping and hunting practices, largely driven by the conservation movement. Growing awareness of the negative impacts of over-hunting and habitat loss led to the establishment of wildlife protection laws and conservation organizations. Notable figures, such as Theodore Roosevelt and John Muir, advocated for the preservation of natural resources and the responsible management of wildlife populations.

In the United States, the Lacey Act of 1900 marked a pivotal moment in wildlife conservation. It prohibited the interstate trade of illegally taken game and provided a framework for regulating hunting practices. Subsequent legislation, such as the Migratory Bird Treaty Act and the Endangered Species Act, further strengthened wildlife protection efforts.

Technological advancements also played a crucial role in modernizing hunting and trapping. The development of firearms, including rifles and shotguns, improved hunting accuracy and efficiency. Traps became more humane, with designs aimed at minimizing injury and stress to captured animals. Additionally, the advent of motor vehicles and communication devices transformed hunting logistics, making it easier to access remote areas and coordinate activities.

Modern Era: Ethical Practices and Sustainable Hunting

In the contemporary era, trapping and hunting have evolved to emphasize ethical practices and sustainability. Modern hunters and trappers are often guided by principles of conservation, respect for wildlife, and ecological balance. Ethical hunting involves fair chase principles, where hunters pursue animals in a manner that does not give them an unfair advantage. This approach fosters a deeper connection with nature and promotes the responsible use of natural resources.

Trapping has also seen significant advancements, with the development of more humane and selective traps. Modern trappers use techniques that prioritize animal welfare, such as live-capture traps that allow for the safe release of non-target species. The use of monitoring systems, such as trail cameras and GPS devices, enhances trap effectiveness while minimizing the impact on wildlife populations.

Furthermore, hunting and trapping have become integral to wildlife management and conservation efforts. Regulated hunting seasons and quotas help control animal populations, preventing overpopulation and habitat degradation. Revenue generated from hunting licenses and permits funds conservation programs, habitat restoration, and research initiatives.

Cultural and Recreational Significance

Hunting and trapping continue to hold cultural and recreational significance for many communities worldwide. Indigenous peoples often maintain traditional hunting practices, passing down knowledge and skills through generations. These practices are not only vital for subsistence but also for preserving cultural heritage and fostering a deep connection to the land.

Recreational hunting and trapping provide opportunities for outdoor enthusiasts to engage with nature, develop survival skills, and contribute to wildlife conservation. Organizations such as Ducks Unlimited and the Rocky Mountain Elk Foundation promote responsible hunting and habitat conservation, ensuring that future generations can enjoy these activities.

Future Perspectives: Challenges and Opportunities

Looking ahead, the future of trapping and hunting faces both challenges and opportunities. Climate change, habitat loss, and human-wildlife conflicts pose significant threats to wildlife populations and ecosystems. Balancing the needs of growing human populations with the conservation of natural resources will require innovative approaches and collaborative efforts.

Advancements in technology, such as drone surveillance and genetic research, offer new tools for wildlife management and conservation. Education and outreach programs can foster a greater appreciation for ethical hunting and trapping practices, encouraging more individuals to become stewards of the environment.

In conclusion, the history and evolution of trapping and hunting reflect humanity's enduring relationship with the natural world. From prehistoric survival to modern conservation, these practices have shaped our cultures, economies, and ecosystems. By embracing ethical and sustainable approaches, we can ensure that trapping and hunting continue to play a positive role in the stewardship of our planet's wildlife and natural resources.

1.2 Ethical Considerations and Legal Regulations

Understanding the ethical considerations and legal regulations surrounding wildlife conservation and trapping is essential for responsible and sustainable practices. These principles ensure that human activities are conducted with respect for wildlife and their habitats, while also complying with laws designed to protect and preserve natural resources.

1. **Ethical Considerations in Trapping and Hunting**

    - **Respect for Wildlife:** Ethical trapping and hunting begin with a profound respect for wildlife. This principle entails recognizing animals as sentient beings that play crucial roles in their ecosystems. Hunters and trappers are encouraged to honor the life of the animal, ensuring that their practices minimize suffering and disruption. This respect extends to the habitats of these animals, emphasizing the need to leave natural environments as undisturbed as possible.
    - **Fair Chase Principles:** The concept of "fair chase" is a cornerstone of ethical hunting. It refers to the pursuit of game in a manner that does not give the hunter an unfair advantage. This principle discourages the use of high-tech gadgets that unduly tip the scales in favor of the hunter, such as drones for spotting game or electronic calls that mimic animal sounds. Instead, fair chase promotes skill, patience, and an intimate understanding of the animal's behavior and habitat.
    - **Minimizing Suffering:** Ethical trappers and hunters prioritize methods that minimize the pain and suffering of animals. In trapping, this involves using traps that are designed to capture animals humanely, such as live-capture or padded leg-hold traps. In hunting, this means ensuring that the hunter is proficient with their weapon of choice, aiming for quick and clean kills. Hunters and

trappers should also be prepared to dispatch wounded animals promptly to prevent prolonged suffering.

- **Selective Harvesting:** Selective harvesting is an ethical practice that involves targeting specific animals based on criteria such as age, sex, and health. This approach helps maintain the balance of animal populations and supports the genetic health of species. For example, hunters might aim to take older, non-breeding animals, allowing younger, healthier individuals to continue reproducing and contributing to the gene pool.
- **Sustainable Practices:** Ethical trapping and hunting are inherently tied to sustainability. This means harvesting animals at levels that do not threaten their populations or disrupt their ecosystems. Hunters and trappers must stay informed about population dynamics and environmental changes to adjust their practices accordingly. Sustainable practices also include habitat conservation efforts, such as participating in or supporting initiatives that protect and restore wildlife habitats.
- **Waste Minimization:** Ethical hunters and trappers ensure that no part of the animal goes to waste. This includes utilizing meat, fur, hides, bones, and other parts for various purposes, whether for personal use, community sharing, or crafting. This practice not only honors the animal's life but also aligns with principles of sustainability and respect for natural resources.

- **Education and Mentorship:** Experienced hunters and trappers have a responsibility to educate and mentor newcomers. This involves passing on ethical practices, skills, and knowledge about wildlife and conservation. By fostering a culture of responsibility and respect, mentors help ensure that future generations of hunters and trappers uphold ethical standards.

Legal Regulations in Trapping and Hunting

- **Historical Development of Regulations:** Legal regulations governing trapping and hunting have evolved significantly over time. Early regulations were often rudimentary, focusing primarily on protecting game populations from over-exploitation. As human impact on wildlife became more pronounced, regulatory frameworks grew more sophisticated, incorporating scientific research and conservation principles.
- **Licensing and Permits:** In most regions, individuals must obtain licenses and permits to legally hunt or trap animals. These documents serve several purposes: they ensure that only qualified individuals engage in these activities, they help regulate the number of participants, and they generate revenue for wildlife conservation programs. Licensing processes often include educational components, such as hunter safety courses, to promote ethical practices and safety.
- **Seasonal Restrictions:** Seasonal restrictions are a common regulatory tool to protect wildlife populations during critical periods, such as breeding and rearing

seasons. By limiting hunting and trapping to specific times of the year, these regulations help ensure that animals have the opportunity to reproduce and raise their young without undue stress or interference. Seasonal restrictions also prevent overharvesting, contributing to the sustainability of wildlife populations.

- **Bag Limits and Quotas:** Bag limits and quotas are established to control the number of animals that can be harvested by an individual or group. These limits are based on scientific assessments of population health and dynamics, ensuring that hunting and trapping activities do not deplete animal numbers. Bag limits can vary by species, region, and season, reflecting the unique conservation needs of different wildlife populations.
- **Protected Species:** Many regions have laws protecting certain species from hunting and trapping due to their endangered status or ecological importance. These protections are crucial for the conservation of biodiversity and the prevention of species extinction. Violating these laws can result in severe penalties, including fines and imprisonment. Hunters and trappers must stay informed about which species are protected and adjust their practices accordingly.
- **Humane Trapping Standards:** Legal regulations often include standards for humane trapping methods. These standards specify the types of traps that can be used, how frequently traps must be checked, and procedures for handling non-target species. For example, many regions prohibit the use of certain

inhumane traps, such as steel-jaw leghold traps, and require trappers to check their traps at least once every 24 hours to minimize animal suffering.

- **Reporting and Monitoring:** Regulatory frameworks typically include requirements for hunters and trappers to report their harvests. This data is crucial for wildlife management, helping authorities monitor population trends and adjust regulations as needed. Reporting systems can be as simple as filling out a tag on a harvested animal or as complex as submitting detailed reports to wildlife agencies. Monitoring also includes compliance checks by wildlife officers to enforce regulations and ensure ethical practices.
- **International Agreements:** Trapping and hunting regulations are not confined to national borders. Many countries are signatories to international agreements that protect migratory species and regulate the trade of wildlife products. For instance, the Convention on International Trade in Endangered Species of Wild Fauna and Flora (CITES) aims to ensure that international trade does not threaten the survival of endangered species. Compliance with these agreements is essential for global wildlife conservation efforts.
- **Public Engagement and Advocacy:** Public engagement and advocacy play vital roles in shaping and enforcing legal regulations. Conservation organizations, hunting and trapping associations, and concerned citizens often work together to advocate for stronger protections and more effective regulations. These efforts

can lead to the passage of new laws, the amendment of existing regulations, and increased funding for conservation programs.

Ethical considerations and legal regulations are deeply intertwined in the practice of trapping and hunting. Ethical practices ensure that hunters and trappers respect wildlife and contribute to the sustainability of animal populations. Legal regulations provide the framework for these practices, ensuring that they are conducted responsibly and with consideration for the broader ecological impact. Together, these elements foster a balanced and respectful relationship between humans and the natural world, promoting conservation and the responsible use of natural resources.

1.3 Importance of Wildlife Conservation

Wildlife conservation is critical for maintaining the health of our planet, ensuring the survival of diverse species, and preserving ecosystems that are vital for human life. It is the practice of protecting wild species and their habitats in order to prevent species from going extinct, reduce the impact of human activities, and ensure that natural resources are available for future generations. Here are several key reasons highlighting the importance of wildlife conservation:

1. Biodiversity and Ecosystem Health

- **Biodiversity as a Foundation:** Biodiversity, the variety of life in all its forms, is the foundation of healthy ecosystems. It includes the diversity of species, genetic variability within those species, and the ecosystems they form. This diversity is essential because it ensures ecosystem resilience, enabling environments to withstand and recover from disturbances like climate change, natural disasters, and human activities. Each species, no matter how small, plays a unique role in maintaining the balance of ecosystems.

- **Ecosystem Services:** Wildlife and biodiversity contribute significantly to ecosystem services, the benefits humans derive from nature. These include provisioning services such as food, water, and raw materials; regulating services like climate regulation, flood control, and disease management; supporting services such as nutrient cycling and soil formation; and cultural services that provide recreational, aesthetic, and spiritual benefits. For instance, bees and other pollinators are critical for the reproduction of many plants, including crops that constitute a significant portion of the human diet.
- **Interconnectedness of Species:** The interconnectedness of species means that the loss of one species can have cascading effects on others. For example, the decline of a predator can lead to an overpopulation of prey species, which in turn can overgraze vegetation and alter the habitat for other species. This interconnected web illustrates the importance of every species in maintaining the balance

and health of ecosystems. Conservation efforts aim to protect this web by preserving species and their habitats.

2. Human Impact and the Need for Conservation

- **Habitat Destruction and Fragmentation:** Human activities such as deforestation, urbanization, and agriculture have led to significant habitat destruction and fragmentation. These activities reduce the available space for wildlife, limit their access to resources, and disrupt migration patterns. Habitat fragmentation can isolate populations, making it difficult for them to find mates and leading to a decrease in genetic diversity. Conservation efforts focus on protecting and restoring habitats to ensure that wildlife populations can thrive.

- **Climate Change:** Climate change poses a significant threat to wildlife. Rising temperatures, changing precipitation patterns, and increasing frequency of extreme weather events alter habitats and ecosystems. Many species are forced to migrate to new areas, adapt to changing conditions, or face extinction. For example, polar bears rely on sea ice for hunting seals, and the melting ice due to global warming threatens their survival. Conservation strategies include efforts to mitigate climate change by reducing greenhouse gas emissions and helping species adapt to changing environments.

- **Overexploitation and Illegal Wildlife Trade:** Overexploitation through activities like hunting, fishing, and logging can deplete wildlife populations to unsustainable levels. Additionally, the illegal wildlife trade poses a significant threat to many species, driven by demand for animal parts for traditional medicine, luxury goods, and exotic pets. Poaching and illegal trade have led to drastic declines in populations of elephants, rhinos, tigers, and many other species. Conservation measures include enforcing laws against illegal trade, creating protected areas, and promoting sustainable use of resources.

- **Pollution:** Pollution, including plastic waste, pesticides, and industrial chemicals, has severe impacts on wildlife. Marine animals often ingest or become entangled in plastic debris, leading to injury or death. Pesticides and chemicals can contaminate soil and water, affecting the health of entire ecosystems. For instance, the pesticide DDT led to the decline of bird species like the bald eagle by thinning their eggshells. Conservation efforts work towards reducing pollution, cleaning up contaminated habitats, and implementing regulations to prevent further damage.

3. Conservation Strategies and Approaches

- **Protected Areas:** Establishing protected areas such as national parks, wildlife reserves, and marine sanctuaries is a critical conservation strategy. These areas provide safe havens for wildlife, free from human exploitation and habitat destruction. They also serve as important sites for scientific research and environmental education. Effective management of protected areas involves monitoring wildlife populations, controlling invasive species, and maintaining natural habitats.

- **Species Recovery Programs:** Species recovery programs focus on the conservation of endangered and threatened species. These programs include captive breeding and reintroduction efforts, habitat restoration, and measures to reduce human-wildlife conflicts. For example, the California condor recovery program involves breeding birds in captivity and releasing them into the wild, coupled with efforts to clean up their habitat and reduce lead poisoning from spent ammunition.
- **Community-Based Conservation:** Engaging local communities in conservation efforts is essential for their success. Community-based conservation programs involve local people in decision-making, benefit-sharing, and active management of natural resources. By providing economic incentives and alternative livelihoods, these programs encourage communities to protect wildlife and their habitats. Examples include ecotourism, sustainable agriculture, and community-managed conservation areas.
- **Policy and Legislation:** Effective conservation requires robust policy and legislation at local, national, and international levels. Laws protecting endangered species, regulating hunting and fishing, and controlling land use are essential tools for conservation. International agreements, such as the Convention on Biological Diversity (CBD) and the Convention on

International Trade in Endangered Species of Wild Fauna and Flora (CITES), play a crucial role in coordinating global conservation efforts. Advocacy and public awareness campaigns are important for promoting and enforcing these laws.
- **Research and Monitoring:** Scientific research and monitoring are fundamental to understanding wildlife populations, their behaviors, and the threats they face. Conservation biologists use various techniques, including field surveys, remote sensing, and genetic analysis, to gather data on species and ecosystems. This information guides conservation strategies and helps evaluate their effectiveness. Monitoring also involves tracking changes in wildlife populations and habitats over time to detect trends and respond to emerging threats.

4. Benefits of Wildlife Conservation

- **Ecological Balance:** Conserving wildlife helps maintain ecological balance and the health of ecosystems. Predators, prey, and other species interact in complex ways that regulate population sizes, control disease, and promote biodiversity. For example, wolves in Yellowstone National Park help control the population of elk, which in turn allows vegetation to recover and benefits other species like beavers and songbirds. Protecting these interactions ensures the stability and functioning of ecosystems.

- **Economic Value:** Wildlife conservation has significant economic benefits. Healthy ecosystems provide resources such as timber, fish, and medicinal plants, which support local economies and livelihoods. Ecotourism, which relies on intact natural environments and abundant wildlife, generates substantial revenue and creates jobs. For instance, wildlife tourism in Africa attracts millions of visitors each year, contributing to the economy and funding conservation efforts.
- **Cultural and Recreational Importance:** Wildlife holds cultural and recreational value for many people around the world. Indigenous communities often have deep spiritual connections to the land and its inhabitants, and traditional practices are closely linked to the health of ecosystems. Recreational activities such as birdwatching, hiking, and wildlife photography provide enjoyment and foster a connection to nature. Conservation efforts help preserve these cultural traditions and recreational opportunities for future generations.
- **Scientific Knowledge and Inspiration:** Wildlife conservation contributes to scientific knowledge and can inspire innovation. Studying wildlife and ecosystems enhances our understanding of biology, ecology, and environmental science. This knowledge can lead to technological advances and solutions to environmental challenges. Additionally, the beauty and complexity of nature have inspired art, literature, and philosophy, enriching human culture and creativity.

5. Global Cooperation and the Future of Conservation

- **International Collaboration:** Global cooperation is essential for addressing the challenges of wildlife conservation. Many species migrate across national borders, requiring coordinated efforts to protect their habitats and regulate human activities. International organizations, treaties, and networks facilitate collaboration among countries, researchers, and conservationists. For example, the Ramsar Convention promotes the conservation of wetlands of international importance, benefiting waterfowl and other species.
- **Adapting to Change:** Conservation strategies must adapt to changing environmental conditions and emerging threats. Climate change, urbanization, and technological advances present new challenges and opportunities for conservation. Adaptive management involves continuously monitoring and adjusting conservation actions based on new information and changing circumstances. This flexible approach helps ensure the long-term success of conservation efforts.
- **Public Engagement and Education:** Engaging the public in conservation is crucial for building support and fostering a conservation ethic. Education programs, outreach activities, and citizen science initiatives raise awareness about the importance of wildlife conservation and encourage people to take action. By connecting individuals to nature and highlighting the benefits of conservation, these efforts help create a culture of stewardship and responsibility.

- **Innovative Approaches:** Innovative approaches and technologies are being developed to enhance wildlife conservation. Conservationists use drones for monitoring wildlife, genetic techniques for species identification, and artificial intelligence for analyzing data. These tools improve the efficiency and effectiveness of conservation efforts. Additionally, approaches like rewilding, which involves restoring ecosystems to their natural state and reintroducing native species, offer new possibilities for conservation.

In conclusion, wildlife conservation is essential for maintaining biodiversity, ecosystem health, and human well-being. It involves a combination of ethical considerations, scientific research, legal regulations, and community engagement. By working together and embracing innovative solutions, we can protect wildlife and their habitats, ensuring a sustainable and thriving planet for future generations.

CHAPTER TWO

Tools and Equipment

2.1 Essential Trapping Tools and Their Uses

Traps

- **Leg-Hold Traps:** Leg-hold traps are among the most common and versatile tools used in trapping. They consist of a pair of spring-powered jaws that snap shut upon the animal's leg when it steps on the trigger plate. These traps are used primarily for capturing furbearers such as foxes, coyotes, and raccoons. Modern leg-hold traps often feature padded jaws to reduce injury to the animal, aligning with ethical trapping standards. Proper setting techniques and trap placement are crucial to ensure the humane capture of the target species while minimizing the risk of capturing non-target animals.

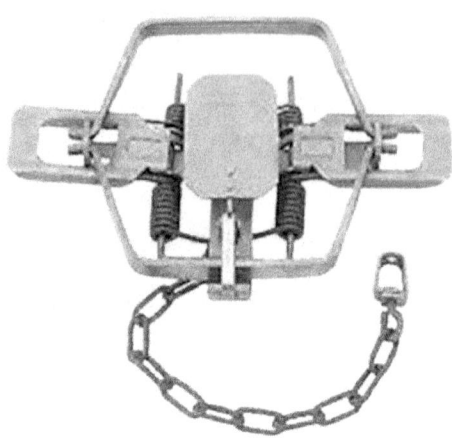

Fig 2: Leg-hold traps

- **Body-Grip Traps:** Body-grip traps, also known as Conibear traps, are designed to kill the animal quickly and humanely. They consist of a square frame with two rotating jaws that close rapidly when triggered, usually around the neck or chest of the animal. These traps are highly effective for aquatic and semi-aquatic species like beavers, muskrats, and otters. Properly setting body-grip traps in runways, dens, or feeding areas maximizes their efficiency and reduces suffering.

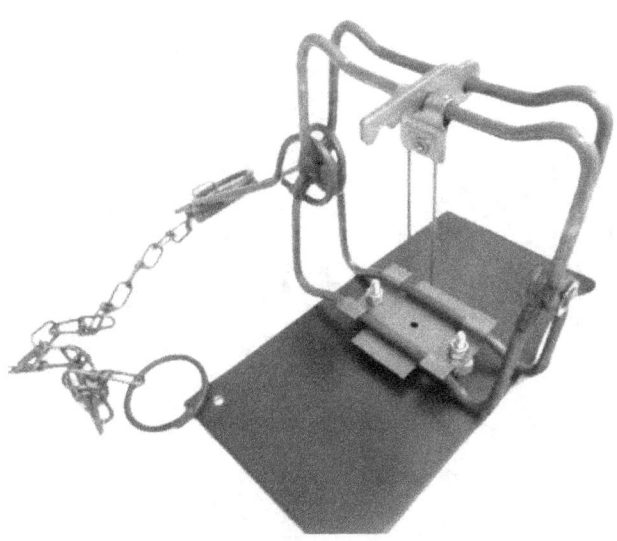

Fig 3: Body-Grip Traps

- **Live-Capture Traps:** Live-capture traps, such as box traps and cage traps, are designed to capture animals without harming them. These traps are ideal for situations where the trapper intends to relocate the animal or release non-target species. They typically feature a baited trigger mechanism that closes the door behind the animal once it enters the trap. Live-capture traps are commonly used for small mammals like rabbits, squirrels, and feral cats, as well as for research purposes and wildlife management.

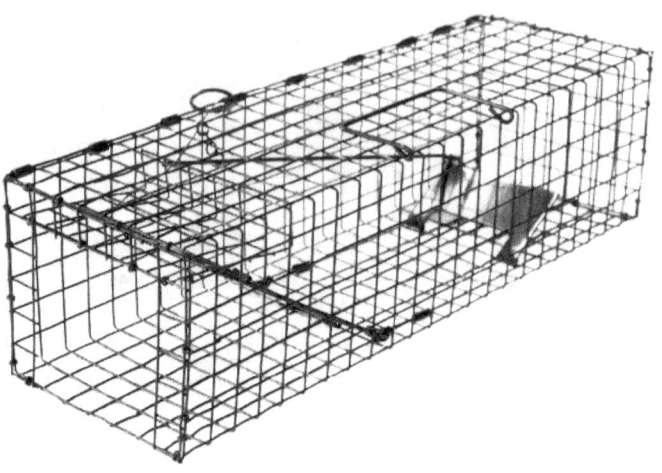

Fig 4: Live-Capture Trap

- **Snares:** Snares are simple, cost-effective traps that consist of a loop of wire or cable that tightens around an animal's body or limb when pulled. They are versatile and can be used for a variety of species, from small game like rabbits to larger animals like deer and boar. Modern snares often include features such as relaxing locks or breakaway devices to minimize injury and allow non-target species to escape. Setting snares requires skill and knowledge of animal behavior to ensure effective and humane captures.

Fig 5: Snares

Tools for Setting and Checking Traps

- **Trap Setters:** Trap setters are tools designed to assist trappers in setting powerful traps, particularly body-grip and large leg-hold traps. These tools provide the leverage needed to compress the springs and position the trap correctly without risking injury to the trapper. Using trap setters enhances efficiency and safety, especially when dealing with strong and potentially dangerous traps.
- **Stake Drivers:** Stake drivers are essential for securing traps in the ground. They are used to drive stakes or anchors into the soil, ensuring that traps remain in place and cannot be dragged away by captured animals. This is particularly important for leg-hold and body-grip traps set for larger or stronger species. Stake drivers come in various sizes and designs to accommodate different types of stakes and soil conditions.

- **Trap Pliers and Wire Cutters:** Trap pliers and wire cutters are multipurpose tools used for adjusting, repairing, and setting traps. They are essential for handling trap chains, swivels, and wire, allowing trappers to make necessary modifications in the field. These tools also facilitate the installation of trap tags, which are often required by law to identify the trapper and ensure compliance with regulations.

Baiting and Luring Equipment

- **Baits and Lures:** Baits and lures are crucial for attracting target species to traps. Baits typically consist of food items that appeal to the animal's diet, such as meat, fish, fruits, or commercial bait formulations. Lures, on the other hand, are scent-based attractants that mimic the natural smells of prey, mates, or territorial markers. Effective use of baits and lures requires knowledge of the target species' feeding habits and scent preferences. Proper bait placement and lure application can significantly increase trapping success rates.
- **Bait Stations and Dispensers:** Bait stations and dispensers are tools used to present bait in a controlled and effective manner. These devices protect bait from weather and non-target animals while ensuring that it remains attractive and accessible to the target species. Bait stations are particularly useful for managing pests in urban or suburban environments, where loose bait could pose risks to pets or children.

- **Lure Bottles and Applicators:** Lure bottles and applicators are designed to dispense scent lures precisely and efficiently. These tools often feature droppers, sprayers, or wicks that allow trappers to apply lures to traps, trails, or specific locations without contaminating their hands or other equipment. Using lure applicators helps maintain the potency and effectiveness of the scent, increasing the likelihood of attracting the target species.

Trapper's Pack and Maintenance Kit

- **Trapper's Pack:** A trapper's pack is an essential piece of equipment for carrying tools, traps, bait, and other supplies in the field. It should be durable, waterproof, and comfortable to carry, with multiple compartments for organizing gear. A well-equipped trapper's pack allows trappers to work efficiently and respond to various situations and conditions encountered during trapping activities.
- **Maintenance Kit:** A maintenance kit includes tools and supplies for repairing and maintaining traps in the field. This typically consists of spare parts like springs, swivels, and jaws, as well as tools such as pliers, wire cutters, and files. Regular maintenance of traps is crucial to ensure their effectiveness and reliability. A well-maintained trap is more likely to function correctly, minimizing the risk of escaped or injured animals.

Field Dressing and Animal Handling Tools

- **Knives and Skinning Tools:** High-quality knives and skinning tools are essential for field dressing and processing captured animals. These tools should be sharp, durable, and easy to handle. Skinning knives with curved blades and gut hooks are particularly useful for efficiently removing hides and preparing animals for further processing. Proper care and maintenance of knives, including regular sharpening and cleaning, are important for ensuring their effectiveness and longevity.
- **Gambrels and Hoists:** Gambrels and hoists are used to suspend animals for skinning and butchering. Gambrels are metal frames that hold the animal's legs apart, providing better access for processing. Hoists, often equipped with pulleys, make it easier to lift and position the animal. These tools are especially useful for larger game, ensuring that the process is hygienic and manageable.

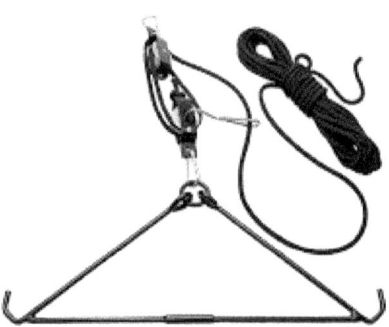

Fig 6: Gambrels and Hoists

- **Protective Gear:** Protective gear, including gloves, aprons, and eye protection, is important for the safety and hygiene of the trapper. Gloves protect hands from cuts, infections, and exposure to animal blood and fluids. Aprons keep clothing clean and provide an additional layer of protection. Eye protection is crucial when working with sharp tools or in situations where there is a risk of splashes or debris.

Navigation and Communication Tools

- **GPS Devices and Maps:** GPS devices and maps are essential for navigation and ensuring that traps are set in optimal locations. GPS devices provide accurate positioning and can help trappers mark and relocate traps efficiently. Topographic maps offer detailed information about terrain and habitat features, aiding in the identification of promising trapping sites. Combining these tools enhances the trapper's ability to navigate unfamiliar areas and maximize trapping success.
- **Two-Way Radios and Mobile Phones:** Two-way radios and mobile phones are important for communication, especially in remote or challenging environments. These tools allow trappers to stay in contact with partners, report their locations, and call for assistance in emergencies. Reliable communication is crucial for safety and coordination during trapping activities, particularly when working in teams or covering large areas.

Recording and Data Management

- **Field Journals and Notebooks:** Field journals and notebooks are used to record important information about trapping activities, including trap locations, dates, bait used, weather conditions, and animal captures. Keeping detailed records helps trappers track their progress, analyze patterns, and make informed decisions. This data is also valuable for reporting requirements and contributing to wildlife management and research efforts.

- **Cameras and Trail Cameras:** Cameras and trail cameras are useful for monitoring trap sites and documenting wildlife activity. Trail cameras, in particular, are motion-activated and can capture images or videos of animals visiting traps. This information helps trappers understand animal behavior, adjust their strategies, and verify the presence of target species. Cameras also provide a non-invasive way to study wildlife and gather evidence for conservation purposes.

A wide range of tools and equipment are essential for effective and ethical trapping. From traps and baiting devices to maintenance kits and protective gear, each piece of equipment plays a crucial role in the trapping process. Understanding the uses and proper handling of these tools ensures that trappers can carry out their activities efficiently, humanely, and in compliance with legal and ethical standards.

2.2 Modern Hunting Gear and Technology

1. Firearms and Ammunition

- **Rifles:** Modern hunting rifles are designed for accuracy, reliability, and versatility. They come in various calibers and configurations to suit different types of game and hunting conditions. Key features to consider include barrel length, stock design, and recoil management. Popular types of hunting rifles include bolt-action, semi-automatic, and lever-action rifles. Each type offers distinct advantages, such as the precision of bolt-action rifles or the rapid-fire capability of semi-automatics.

Fig 7: Rifle

- **Shotguns:** Shotguns are used for hunting birds and small game, such as waterfowl and upland birds. They are available in various gauges, with 12-gauge and 20-gauge being the most common. Modern shotguns feature interchangeable choke tubes, which allow hunters to adjust the spread of the shot pattern to match different hunting scenarios. Pump-action, semi-automatic, and break-action are common actions for shotguns, each offering different benefits in terms of speed and ease of use.

Fig 8: Shotgun

- **Ammunition:** Modern ammunition is engineered for performance and precision. Choices include various types of bullets and shot loads, such as hollow points, full metal jackets, and lead or steel shot. Selecting the appropriate ammunition for the target species and hunting conditions is crucial for ensuring effective and humane kills. Advances in ammunition technology, such as bonded bullets and advanced shot materials, enhance accuracy, expansion, and terminal performance.

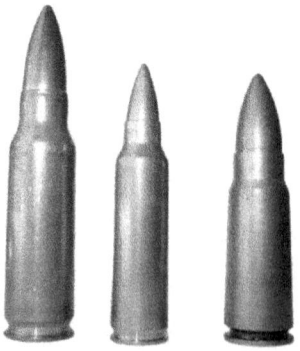

Fig 9: Ammunition

2. Optics and Sights

- **Scopes:** Rifle scopes provide magnification and improved accuracy for long-range shots. Modern scopes feature variable magnification, adjustable objective lenses, and illuminated reticles. Features such as parallax adjustment, turrets for elevation and windage, and reticle options tailored for different game and conditions enhance precision and shooting confidence. Choosing the right scope depends on the intended hunting distance, target size, and personal preference.
- **Binoculars:** Binoculars are essential for scouting and spotting game. Modern binoculars offer high magnification, wide fields of view, and durable, weather-resistant designs. Features such as image stabilization and low-light performance enhance their effectiveness in various hunting environments. Optics quality, including lens coatings and prisms, affects brightness, clarity, and contrast.
- **Rangefinders:** Rangefinders measure the distance between the hunter and the target, aiding in shot accuracy. They use laser technology to provide precise distance readings and often include features such as angle compensation and ballistic calculators. Rangefinders help hunters make informed decisions about shot placement and trajectory, improving their chances of a successful and ethical hunt.

3. Clothing and Gear

- **Camouflage Clothing:** Camouflage clothing is designed to blend the hunter into their environment, reducing the likelihood of detection by game. Modern camouflage patterns are tailored for specific habitats, such as forest, grassland, or marshland. Clothing options include jackets, pants, gloves, and hats, often made from lightweight, breathable, and water-resistant materials. Layering systems provide versatility and comfort in varying weather conditions.
- **Hunting Boots:** Hunting boots are designed for durability, comfort, and protection in rugged terrain. Features to consider include waterproofing, insulation, ankle support, and tread patterns for traction. Modern boots often incorporate advanced materials and technologies, such as Gore-Tex linings and Vibram soles, to enhance performance and comfort during long hunts.
- **Hunters' Safety Gear:** Safety gear is crucial for preventing accidents and ensuring hunter safety. Essential items include blaze orange clothing to enhance visibility and prevent accidental shootings, as well as safety harnesses for elevated stands or blinds. Other safety gear may include first aid kits, multi-tools, and emergency signaling devices. Proper safety measures are fundamental to a safe and enjoyable hunting experience.

4. Navigation and Communication Tools

- **GPS Devices:** GPS devices are essential for navigation and tracking during hunts. Modern handheld GPS units offer detailed maps, waypoint marking, and tracking capabilities. Some devices also include features like compass functions, altimeters, and weather forecasts. GPS technology helps hunters navigate unfamiliar terrain, mark important locations, and track their routes for efficient and safe hunting.

- **Smartphones and Apps:** Smartphones equipped with hunting and navigation apps provide additional functionality and convenience. Apps can offer features such as mapping, weather updates, game calling, and hunting regulations. Many apps also include social networking elements, allowing hunters to share experiences and tips with others. Smartphones also serve as communication tools for staying in touch with hunting partners and accessing emergency services if needed.

5. Hunting Accessories

- **Calls and Decoys:** Calls and decoys are used to attract game and improve hunting success. Calls mimic the sounds of prey or mating calls, drawing animals into shooting range. Various types of calls, such as turkey calls, duck calls, and predator calls, are available, each designed to replicate specific sounds. Decoys, which can be lifelike models of animals or birds, are used to create realistic scenarios and attract game by simulating the presence of other animals.
- **Game Cameras:** Game cameras, also known as trail cameras, are used to monitor wildlife activity and patterns. These cameras are equipped with motion

sensors and infrared technology to capture images or videos of animals in their natural habitat. Game cameras help hunters identify active areas, understand animal behavior, and plan their hunting strategies. They are also valuable tools for wildlife research and management.

- **Scent Control Products:** Scent control products help hunters reduce their odor and minimize detection by game. These products include scent-eliminating sprays, specialized hunting clothing, and odor-neutralizing soaps. Some hunters also use scent-dispensing devices to simulate natural scents or cover their own scent. Effective scent control enhances the chances of a successful hunt by making it more difficult for animals to detect human presence.

6. Advanced Technology and Innovations

- **Thermal and Night Vision Scopes:** Thermal and night vision scopes enable hunting in low-light or nighttime conditions. Thermal scopes detect heat signatures and create images based on temperature differences, making them useful for detecting game in complete darkness. Night vision scopes amplify available light to provide clear images in low-light conditions. These technologies are particularly beneficial for hunting nocturnal animals or for safety during early morning or late evening hunts.
- **Smart Rifles and Ammo:** Smart rifles and ammunition incorporate advanced technologies to enhance shooting accuracy and performance. Some

smart rifles feature built-in ballistic computers and electronic triggers that adjust for distance, wind, and other factors. Smart ammunition may include sensors or programmable elements that optimize terminal performance. These innovations offer increased precision and efficiency, although they may be subject to regulatory restrictions in some regions.

- **Drones:** Drones are used for scouting and monitoring hunting areas. Equipped with cameras, drones can survey large areas, identify game locations, and assess terrain features. They provide valuable aerial perspectives and can be used to gather information without disturbing wildlife. However, the use of drones for hunting is regulated in many areas, and hunters must ensure compliance with local laws and ethical guidelines.
- **Automated Game Feeders:** Automated game feeders dispense food at scheduled intervals, attracting game to specific locations. These feeders can be set up to provide consistent and reliable food sources, helping hunters pattern animal movements and increase their chances of successful hunts. Automated feeders are often used in conjunction with game cameras to monitor and analyze wildlife activity.

7. Maintenance and Care Tools

- **Cleaning Kits:** Cleaning kits are essential for maintaining firearms and ensuring their proper function. These kits typically include brushes, rods, patches, and solvents for cleaning barrels, chambers, and

other components. Regular cleaning and maintenance prevent malfunctions, ensure accuracy, and prolong the lifespan of hunting equipment.
- **Gun Cases and Storage:** Gun cases and storage solutions protect firearms from damage and ensure their safe transport. Cases are available in various styles, such as hard cases and soft cases, with features like padded interiors and locking mechanisms. Proper storage and handling are crucial for maintaining the integrity of hunting gear and ensuring safety.

Modern hunting gear and technology enhance the efficiency, accuracy, and enjoyment of hunting. From firearms and optics to navigation tools and advanced technologies, each piece of equipment plays a crucial role in successful and ethical hunting practices. Understanding and properly utilizing these tools ensures a more effective and enjoyable hunting experience, while also contributing to safety and conservation efforts.

2.3 Maintenance and Safety of Equipment

Maintaining and ensuring the safety of trapping and hunting equipment is crucial for effective and ethical practices. Proper care extends the life of your gear, enhances performance, and promotes safety for both the user and wildlife. This section provides comprehensive guidelines on maintaining and safely using trapping and hunting equipment.

General Maintenance Practices

- **Regular Inspection:** Regular inspection of hunting equipment is crucial to ensure its functionality and safety. Before each hunting trip, thoroughly check all gear, including firearms, traps, and clothing. Look for signs of wear, damage, or malfunction. For firearms, inspect the barrel, action, and safety mechanisms. For traps, check the springs, jaws, and triggers. Early detection of issues can prevent accidents and ensure equipment performs as expected.
- **Cleaning:** Proper cleaning is essential for maintaining equipment and ensuring its longevity. Firearms should be cleaned after each use to remove residue, moisture, and dirt. Use appropriate solvents, brushes, and patches to clean the barrel, action, and chamber. For traps, clean dirt, rust, and debris from moving parts and ensure they function smoothly. Hunting clothing and boots should be washed according to manufacturer instructions to maintain their effectiveness and durability.
- **Lubrication:** Lubrication helps maintain the smooth operation of moving parts and prevents rust and corrosion. Use suitable lubricants for each type of equipment. For firearms, apply oil to the moving parts and ensure it is evenly distributed. Traps may require occasional lubrication on the springs and swivels to ensure proper function. Avoid over-lubricating, as excess oil can attract dirt and debris.
- **Storage:** Proper storage is vital to protect equipment from damage and environmental factors. Store firearms in a secure, dry location, preferably in a locked cabinet or

safe. Use gun cases or sleeves to protect them from scratches and moisture. Traps should be stored in a dry, cool place to prevent rust and deterioration. Hunting clothing and gear should be cleaned and stored in a dry, well-ventilated area to prevent mold and mildew growth.

Firearms Maintenance

- **Barrel Care:** The barrel is a critical component of a firearm and requires regular cleaning to maintain accuracy and safety. Use a cleaning rod with a bore brush to remove fouling and debris from the barrel. Follow up with patches and solvent to clean the bore thoroughly. Regular cleaning prevents buildup that can affect performance and safety.
- **Action and Chamber Cleaning:** The action and chamber of a firearm should be cleaned to remove residue and ensure smooth operation. Use a brush or cleaning tool to remove dirt and carbon buildup from these areas. Apply a light coat of oil to moving parts to prevent rust and ensure proper functioning.
- **Safety Checks:** Regularly check the safety mechanisms of your firearm to ensure they function correctly. Test the safety lever, trigger, and any other safety features to verify that they engage and disengage properly. A malfunctioning safety can pose serious risks, so address any issues promptly.

Traps Maintenance

- **Rust Prevention:** Traps are exposed to various environmental conditions, which can lead to rust and corrosion. To prevent rust, clean traps after each use and apply a protective coating, such as rust-resistant paint or oil. Store traps in a dry place to minimize moisture exposure.
- **Spring and Jaw Inspection:** Regularly inspect the springs and jaws of traps for signs of wear or damage. Ensure that the springs have sufficient tension and that the jaws close properly. Replace any damaged parts to maintain the trap's effectiveness and safety.
- **Trigger Mechanism Care:** The trigger mechanism is a critical component of traps and should be checked for proper function. Ensure that the trigger is sensitive and responsive. Clean and lubricate the trigger mechanism as needed to maintain its performance.

Clothing and Footwear Care

- **Cleaning and Washing:** Follow the manufacturer's instructions for cleaning hunting clothing and footwear. Use appropriate detergents and avoid fabric softeners, which can reduce the effectiveness of camouflage and waterproofing treatments. Wash hunting clothing separately to avoid contaminating other items with scents or residues.
- **Waterproofing:** Many hunting clothing and boots are treated with waterproofing agents. Reapply these treatments periodically to maintain water resistance. Use

specialized waterproofing sprays or treatments designed for the specific materials of your gear.
- **Repair and Replacement:** Inspect hunting clothing and boots for signs of wear and damage. Repair minor issues, such as small tears or broken zippers, promptly to extend the life of the gear. Replace items that are excessively worn or damaged to ensure safety and performance.

Safety Practices

- **Proper Handling:** Always handle hunting equipment with care and respect. For firearms, treat every gun as if it is loaded and keep the muzzle pointed in a safe direction. Follow all safety protocols when handling traps, including wearing gloves and using appropriate tools.
- **Training and Familiarization:** Ensure that you are thoroughly familiar with the operation and maintenance of all hunting equipment. Attend training courses or seek guidance from experienced hunters if needed. Proper knowledge and training reduce the risk of accidents and improve your effectiveness as a hunter.
- **Emergency Preparedness:** Be prepared for emergencies by carrying a first aid kit and knowing basic first aid procedures. Familiarize yourself with emergency protocols for dealing with accidents or injuries. Have a plan for contacting emergency services and ensure that your hunting partners are aware of it.

- **Compliance with Regulations:** Adhere to all local and national regulations regarding hunting equipment and safety. This includes using approved gear, following legal hunting practices, and complying with safety guidelines. Understanding and following regulations helps ensure a safe and responsible hunting experience.

Electronic Devices

- **Battery Maintenance:** For electronic devices such as GPS units, rangefinders, and game cameras, regularly check and replace batteries to ensure they function properly during hunting trips. Store spare batteries in a dry, cool place to maintain their effectiveness.
- **Device Care:** Keep electronic devices clean and dry. Use protective cases or covers to shield them from environmental elements. Follow the manufacturer's instructions for maintenance and storage to prolong the lifespan of the devices.

Maintaining and ensuring the safety of hunting equipment involves regular inspection, cleaning, lubrication, and proper storage. By following these practices, hunters can ensure that their gear remains functional, safe, and effective. Adhering to safety protocols and regulations is essential for a successful and responsible hunting experience.

CHAPTER THREE

Trapping Techniques and Methods

3.1 Types of Traps and Their Applications

Trapping is a versatile and effective method for capturing wildlife, and different types of traps are designed for specific purposes and species. Understanding the various types of traps and their applications is essential for successful and ethical trapping. Below is an overview of the main types of traps and their uses:

1. Leg-Hold Traps

Leg-hold traps consist of a pair of spring-loaded jaws that snap shut when triggered by the animal's movement. They are typically made from metal and come in various sizes to accommodate different species.

Applications:

- **Furbearers:** Commonly used for capturing animals such as coyotes, foxes, raccoons, and beavers. Their design allows for effective and humane capture if set and used correctly.

- **Padded Jaws:** Modern leg-hold traps often feature padded jaws to minimize injury and discomfort, aligning with ethical trapping standards.

- **Setting:** Traps are set along animal trails, near dens, or in feeding areas. Proper placement is crucial to ensure effectiveness and reduce the risk of catching non-target species.

2. Body-Grip Traps (Conibear Traps)

Body-grip traps, also known as Conibear traps, are designed to kill the animal quickly by crushing its body between two powerful, spring-loaded jaws. They are typically used in a variety of sizes, from small to large, depending on the target species.

Applications:

- **Aquatic and Semi-Aquatic Species:** Effective for trapping beavers, muskrats, and otters, as well as land-dwelling animals like raccoons.

- **Efficiency:** The traps are set in runways, dens, or along water channels. They are designed to kill quickly and humanely, but careful placement is needed to avoid non-target captures.

- **Regulations:** Some regions have specific regulations on the use of body-grip traps due to concerns about their impact on non-target species.

3. Live-Capture Traps

Live-capture traps, including box traps and cage traps, are designed to capture animals without causing harm. They consist of a cage or box with a door that closes when the animal enters.

Applications:

- **Small Mammals:** Ideal for capturing animals such as rabbits, squirrels, and feral cats. They are also used for relocating animals or for research purposes.
- **Non-Harmful:** These traps are used when the goal is to relocate the animal or release it back into the wild.
- **Baiting:** Live-capture traps are typically baited with food or scent attractants to entice the animal inside.

4. Snares

Snares are simple traps made from a loop of wire or cable that tightens around the animal's body or limb when triggered. They are often used in a variety of settings, including trails and dens.

Applications:

- **Versatile Use:** Suitable for capturing a range of species from small game like rabbits to larger animals such as deer or boar.

- **Adjustable:** Snares can be adjusted to fit different sizes of animals and can be used for both lethal and live capture, depending on the design.

- **Setting:** Snares require skillful placement and knowledge of animal behavior to ensure effectiveness and to avoid catching non-target species.

5. Foot Hold Traps

Foot hold traps are similar to leg-hold traps but may vary in design and mechanism. They are used to capture an animal by its foot, often with a mechanism that prevents escape.

Applications:

- **Furbearers:** Often used for trapping animals like bobcats, otters, and coyotes. These traps are designed to hold the animal securely until the trapper arrives.

- **Specialized Designs:** Some foot hold traps have features like offset jaws or rubber padding to reduce injury and discomfort.

- **Placement:** Set along trails, in dens, or near feeding areas to maximize effectiveness.

6. Kill Traps

Kill traps are designed to kill the animal instantly upon capture. They include various mechanisms such as spring-loaded jaws or blade systems.

Applications:

- **Quick and Humane:** Ideal for situations where a quick, humane kill is desired. Commonly used for rodents and small to medium-sized animals.
- **Types:** Include various designs, such as the rat trap and larger models for catching species like weasels or minks.
- **Regulations:** Some areas have specific rules regarding the use of kill traps to ensure humane treatment and prevent unintended harm to non-target species.

7. Box Traps

Box traps are a type of live-capture trap with a rectangular box or cage that closes when the animal enters. They often have a spring-loaded door that snaps shut upon activation.

Applications:

- **Small to Medium Mammals:** Suitable for capturing animals like squirrels, opossums, and stray cats.
- **Ease of Use:** Simple to set and use, with the ability to monitor and relocate animals easily.
- **Baiting:** Box traps are baited with food or scent attractants to lure animals inside.

8. Foothold Traps

Foothold traps are designed to capture animals by their foot. They are similar to leg-hold traps but may have different jaw configurations and trigger mechanisms.

Applications:

- **Large and Small Game:** Used for trapping various species, including coyotes, foxes, and bobcats. Foothold traps are often preferred for their ability to hold animals securely without causing excessive harm.
- **Adjustable:** Various sizes and designs are available to suit different species and trapping conditions.
- **Regulations:** Regulations may vary regarding the use of foothold traps, with some areas requiring specific features to ensure humane treatment.

9. Pitfall Traps

Pitfall traps consist of a hole or pit that the animal falls into when it steps over the edge. These traps are usually used in conjunction with bait or scent attractants.

Fig 10: Pitfall Traps

Applications:

- **Ground-Dwelling Species:** Suitable for capturing animals like small mammals or insects that travel along the ground.

- **Temporary Use:** Often used for short-term trapping and research purposes. Pitfall traps can be effective for capturing a range of species but require regular monitoring to ensure the trapped animals are not left in distress.

10. Mousetraps

Mousetraps are small traps designed specifically for capturing or killing mice. They come in various designs, including snap traps and live traps.

Fig 11: Mousetraps

Applications:

- **Rodents:** Effective for managing rodent populations in homes, barns, and other structures. Mousetraps are designed for quick and effective rodent control.

- **Types:** Include snap traps for killing and live traps for capturing and relocating mice.

Different types of traps serve various purposes and are suited for capturing different species. Understanding the specific applications, setting techniques, and ethical considerations for each type of trap is crucial for successful and responsible trapping. By selecting the appropriate trap for the target species and following best practices, trappers can achieve their goals while minimizing harm to non-target animals and maintaining compliance with regulations.

3.2 Setting and Baiting Traps Effectively

Setting and baiting traps are critical skills in successful trapping. The effectiveness of a trap largely depends on how well it is set and the attractiveness of the bait used. Here's an in-depth guide on how to set and bait traps effectively:

1. Understanding Animal Behavior

i. Species-Specific Behavior:

- **Habitat Preferences:** Different animals have unique habitat preferences. Understanding these preferences helps in choosing the right location for your trap.

- **Feeding Habits:** Knowing what an animal eats will guide you in selecting the appropriate bait. For example, raccoons are attracted to sugary or fatty foods, while foxes may be drawn to meat-based baits.

- **Movement Patterns:** Observing the animal's movement patterns helps in placing the trap in high-traffic areas, such as trails or near dens.

ii. Scent and Sight Cues:

- Animals rely heavily on their sense of smell and sight. Properly using scents and visual attractants can increase the likelihood of a successful capture.

2. Choosing the Right Location

i. High-Traffic Areas:

- **Animal Trails:** Set traps along established animal trails, where animals frequently travel.

- **Feeding Areas:** Place traps near food sources or feeding areas, as animals are more likely to encounter the trap while foraging.

ii. **Shelter and Cover:**

- **Dens and Nests:** Position traps near dens, nests, or burrows to catch animals when they are coming or going.

- **Natural Cover:** Use natural cover, such as rocks or vegetation, to conceal the trap and make it less conspicuous to the animal.

iii. **Environmental Conditions:**

- **Weather Considerations:** Avoid setting traps in areas prone to extreme weather conditions, as this can affect the trap's functionality and the effectiveness of the bait.

3. Setting Different Types of Traps

i. **Leg-Hold Traps:**

- **Preparation:** Flatten the ground where the trap will be set. Ensure that the trap is stable and won't move when an animal steps on it.

- **Placement:** Position the trap so that the animal's foot will be caught by the trap when it steps on the pan. This often involves placing the trap directly in the animal's path.

- **Setting Mechanism:** Follow the manufacturer's instructions to set the trap properly. Ensure that the spring mechanism is functioning correctly.

ii. **Body-Grip Traps:**

- **Positioning:** Place the trap in a natural funnel or runway where the animal is likely to pass through.

- **Setting:** Make sure the trap is securely anchored and aligned with the funnel to ensure a clean and effective capture.

- **Safety:** Always handle body-grip traps with care, as they are powerful and can cause injury.

iii. **Live-Capture Traps:**

- **Baiting:** Place bait inside the trap to attract the animal. Ensure that the bait is securely positioned and accessible to the animal.

- **Trigger Mechanism:** Check the trap's trigger mechanism to ensure it will activate when the animal enters the trap.

- **Monitoring:** Regularly check live-capture traps to avoid stress or harm to the trapped animal.

iv. **Snares:**

- **Loop Size:** Adjust the size of the snare loop to match the size of the target animal. The loop should be large enough to capture the animal but not so large that it misses.

- **Setting:** Position the snare where the animal is likely to pass through, such as trails or near dens. Ensure that the snare is properly anchored and won't move.

v. Box Traps:

- **Bait Placement:** Place the bait at the far end of the box to encourage the animal to fully enter the trap.

- **Setting:** Ensure that the door mechanism is properly set and will close securely when the animal enters.

4. Choosing and Using Bait

i. Types of Bait:

- **Food Bait:** Use food items that are appealing to the target species. Examples include meat, fish, fruits, or grains. For example, bacon or fish for raccoons, and rabbit or squirrel meat for foxes.

- **Scent Bait:** Scent attractants, such as pheromones or urine, can lure animals. These are especially useful for predators or species with strong scent-marking behaviors.

ii. Bait Placement:

- **Inside Traps:** Place bait inside the trap to draw the animal in. Ensure that the bait is positioned in a way that encourages the animal to trigger the trap.

- **Around Traps:** For some traps, placing bait around the trap can attract the animal to the trap area. This is particularly useful for setting snares or body-grip traps.

iii. Bait Freshness:

- **Regular Replacement:** Replace bait regularly to ensure it remains fresh and attractive to animals. Spoiled or old bait may lose its effectiveness.

- **Proper Storage:** Store bait properly to prevent contamination and spoilage. Use airtight containers for perishable baits and keep them in a cool, dry place.

5. Monitoring and Maintenance

i. Regular Checks:

- **Frequency:** Check traps frequently to ensure they are functioning correctly and to minimize the time an animal spends in the trap.

- **Adjustments:** Make necessary adjustments to traps or bait based on the effectiveness and any signs of non-target captures or damage.

6. Safety Precautions

i. Handling Traps:

- **Protective Gear:** Wear gloves and other protective gear when handling traps to prevent injury and avoid transferring human scent.

- **Safe Setting:** Ensure that traps are set in a safe manner, particularly when using powerful or potentially dangerous traps like body-grip traps.

Effective setting and baiting of traps require a thorough understanding of animal behavior, proper placement and adjustment of traps, and the use of appropriate bait. By carefully selecting trap types, baiting strategies, and monitoring practices, trappers can enhance their success while adhering to ethical and legal standards. Regular maintenance, safety precautions, and compliance with regulations are essential for responsible and effective trapping.

3.3 Monitoring and Checking Traps

Monitoring and checking traps are vital practices for ensuring humane treatment of trapped animals, maintaining the effectiveness of trapping efforts, and complying with legal and ethical standards. Here's an extensive guide on how to monitor and check traps effectively:

1. Frequency of Checks

i. Regulatory Requirements:

- **Local Regulations:** Most regions have specific regulations that dictate how frequently traps must be checked. These regulations are designed to ensure the humane treatment of trapped animals and to minimize their suffering.

- **Common Intervals:** Typical check intervals range from every 24 to 72 hours, depending on local laws and the type of trap used. For instance, live traps may require more frequent checks compared to body-grip traps.

ii. Practical Considerations:

- **Type of Trap:** The frequency of checks can depend on the trap type and its impact on the captured animal. For live traps and traps that can potentially cause harm, more frequent checks are crucial.

- **Environmental Conditions:** Weather conditions and the presence of predators or scavengers can influence how often traps need to be checked. In harsh weather or high predator areas, more frequent checks may be necessary.

2. Inspection Process

i. Visual Inspection:

- **Check Placement:** Ensure that the trap is still properly placed and hasn't been displaced by animals, weather, or other factors.

- **Examine Trap Condition:** Look for any signs of damage, wear, or malfunction in the trap mechanism. For example, check that the springs and jaws of leg-hold traps are functioning correctly.

ii. Animal Welfare:

- **Assess Animal Condition:** If an animal is trapped, check its condition to ensure it is not suffering excessively. Look for signs of stress, injury, or illness.

- **Humane Treatment:** Handle the trapped animal with care to minimize stress and injury. If it's a live trap, release the animal promptly in accordance with local regulations and ethical guidelines.

iii. Bait and Setting:

- **Bait Status:** Check if the bait has been consumed or is still fresh. Replenish or replace bait as necessary to maintain its attractiveness.

- **Trap Setting:** Ensure that the trap is still set correctly and has not been tampered with or damaged.

3. Handling Trapped Animals

i. Safety Precautions:

- **Protective Gear:** Wear gloves and other protective equipment when handling traps and animals to prevent injury and avoid transferring human scent.

- **Avoid Direct Contact:** Minimize direct contact with the animal to reduce stress and potential harm. Use tools or equipment to handle the trap and the animal if needed.

ii. Humane Release:

- **Prompt Action:** Release live-trapped animals as soon as possible. Ensure the release site is suitable for the species and complies with local regulations.

- **Transport:** If transporting a trapped animal to a different location, do so in a manner that minimizes stress and discomfort. Use a covered container for transport and avoid exposing the animal to extreme temperatures or handling.

4. Documentation and Record Keeping

i. Record Keeping:

- **Trap Locations:** Maintain records of trap locations, settings, and bait used. This information helps in evaluating trap effectiveness and making necessary adjustments.

- **Catch Records:** Document details about the animals captured, including species, size, and condition. This data is useful for monitoring trends, assessing trap performance, and complying with reporting requirements.

ii. Monitoring Logs:

- **Check Intervals:** Keep a log of when each trap is checked and any observations made during the inspection. This helps ensure compliance with regulations and provides a record for review and analysis.

- **Adjustments:** Note any adjustments made to traps or bait and their outcomes. This information can help refine trapping strategies and improve results.

5. Emergency Situations

i. Handling Injuries:

- **Injured Animals:** If a trapped animal is injured or in distress, assess its condition and decide on the appropriate action. In some cases, it may be necessary to contact a wildlife rehabilitator or animal control officer.

- **Trap Malfunctions:** If a trap malfunctions or poses a risk to the animal, address the issue immediately to prevent further harm. Repair or replace faulty traps as needed.

ii. Safety Procedures:

- **Emergency Contacts:** Have contact information for local wildlife authorities or animal control officers in case of emergencies. This is important for addressing situations involving injured or problematic animals.

- **First Aid:** Be prepared to administer first aid to yourself or others in case of accidents or injuries while checking or handling traps. Carry a basic first aid kit and know how to use its contents.

Monitoring and checking traps effectively involve adhering to legal requirements, conducting thorough inspections, ensuring humane treatment of animals, and maintaining detailed records. Regular checks, proper handling, and ethical practices are essential for successful trapping and the well-being of captured animals. By following best practices and addressing any issues promptly, trappers can ensure a responsible and effective trapping operation.

CHAPTER FOUR

Modern Hunting Strategies

4.1 Tracking and Stalking Techniques

Tracking and stalking are fundamental skills in hunting and trapping, allowing hunters to locate, follow, and approach their target animals effectively. Mastery of these techniques involves understanding animal behavior, reading signs, and using stealth and strategy. Here's an in-depth guide on tracking and stalking techniques:

1. Understanding Animal Behavior

i. Species-Specific Knowledge:

- **Movement Patterns:** Different animals have distinct movement patterns based on their daily routines, feeding habits, and environmental preferences. For instance, deer are typically most active during dawn and dusk, while predators like coyotes may be more active at night.

- **Habitat Preferences:** Knowing where an animal prefers to live helps in finding their tracks and signs. For example, beavers are often found near water bodies, while upland birds prefer forested or grassland areas.

ii. Behavioral Indicators:

- **Feeding Habits:** Understanding what and where an animal feeds can help in identifying feeding grounds or trails. For instance, elk often graze in meadows, while squirrels may be found in areas with abundant nut-bearing trees.

- **Social Structures:** Some animals have social structures that influence their behavior. For example, wolves travel in packs, and tracking their movements might involve following multiple sets of tracks.

2. Reading Tracks and Signs

i. Track Identification:

- **Footprints:** Tracks provide valuable information about an animal's size, weight, and gait. Study the shape, size, and depth of footprints to identify the species and determine its direction of travel.

- **Gait Patterns:** Different animals have distinctive gait patterns, such as walking, trotting, or running. Understanding these patterns helps in interpreting the tracks and estimating the animal's speed.

ii. Signs of Activity:

- **Droppings:** Animal droppings can reveal dietary habits, health status, and recent activity. Fresh droppings are more likely to indicate recent presence, while older droppings may suggest past activity.

- **Scratches and Rubs:** Scratches on trees, rubs on shrubs, or broken branches can indicate an animal's presence and behavior. For example, deer may rub their antlers on trees to mark territory.

iii. Other Indicators:

- **Nests and Burrows:** Nests, burrows, and dens indicate areas where animals live or rest. Examining these structures can provide insights into the animal's habits and movements.

- **Tracks in Mud or Snow:** Tracking is often easier in mud or snow, where tracks are more clearly defined. Look for disturbances in the ground, such as scuffed areas or disturbed vegetation.

3. Tracking Techniques

i. Initial Assessment:

- **Identify Tracks:** Start by identifying the type of animal based on the tracks. Look for distinctive features such as claw marks, pad shapes, and track size.

- **Direction of Travel:** Determine the direction in which the animal is moving by following the tracks. This helps in planning your approach and understanding the animal's route.

ii. Following Tracks:

- **Consistency:** Follow the tracks consistently, noting changes in direction or behavior. If the track pattern changes, the animal may have altered its path or encountered obstacles.

- **Use of Tools:** Use tracking tools, such as track plates or trail cameras, to gather more information about the animal's movements and habits.

iii. Estimating Time:

- **Track Freshness:** Assess the freshness of the tracks to estimate when the animal passed through the area. Fresh tracks are softer and more defined, while older tracks may be weathered or less distinct.

- **Environmental Conditions:** Consider environmental factors, such as rain or snow, which can affect the appearance of tracks and signs. Adjust your tracking strategy accordingly.

4. Stalking Techniques

i. Approach Strategy:

- **Wind Direction:** Always approach from downwind to avoid alerting the animal with your scent. Understanding wind direction helps in maintaining stealth and reducing the chance of detection.

- **Cover and Concealment:** Use natural cover, such as trees, bushes, and terrain features, to conceal your presence. Move slowly and stay low to the ground to avoid being seen.

ii. Movement and Stealth:

- **Slow and Steady:** Move slowly and deliberately, making minimal noise. Avoid sudden movements that could startle the animal or draw attention.

- **Foot Placement:** Watch your foot placement to avoid stepping on noisy or brittle materials. Move on soft ground or use padded footwear to minimize noise.

iii. Sound and Scent Management:

- **Control Noise:** Be mindful of any sounds you make, such as rustling clothing or breaking twigs. Use quiet, deliberate movements to reduce noise.

- **Scent Management:** Avoid strong odors, such as perfumes or food smells, which can alert animals to your presence. Use scent-masking products or natural cover scents if necessary.

iv. Visual Observation:

- **Scan the Area:** Continuously scan the area for visual cues, such as movement or color changes. Look for signs of the animal's presence, such as fur, feathers, or other markings.

- **Use Binoculars:** Binoculars can help you spot animals from a distance without disturbing them. Use them to get a closer look at distant animals or to confirm their presence.

5. Handling and Adjusting to Changing Conditions

i. Environmental Changes:

- **Weather Conditions:** Be prepared to adapt to changing weather conditions, such as rain, snow, or wind, which can affect tracking and stalking efforts. Adjust your techniques based on visibility and track clarity.

- **Terrain Variations:** Different terrains present different challenges for tracking and stalking. Adapt your approach based on the terrain, whether it's dense forest, open fields, or rocky areas.

ii. Animal Behavior Changes:

- **Adaptation:** Animals may alter their behavior in response to environmental changes or increased human activity. Be prepared to adjust your tracking and stalking strategies as needed.

- **Learning from Experience:** Use your observations and experiences to refine your techniques. Learn from each tracking and stalking experience to improve your skills and effectiveness.

Tracking and stalking are essential skills for effective hunting and trapping. Mastering these techniques involves understanding animal behavior, reading tracks and signs, employing stealth and strategy, and adapting to changing conditions. By honing these skills and adhering to ethical and legal standards, hunters and trappers can achieve successful outcomes while respecting wildlife and their habitats.

4.2 Using Calls and Decoys

Calls and decoys are powerful tools for attracting and deceiving wildlife, enhancing the effectiveness of hunting and trapping efforts. By mimicking animal sounds or creating visual representations, hunters can draw animals into range, making it easier to observe or capture them. Here's an in-depth guide on using calls and decoys effectively:

1. Understanding Calls and Their Uses

i. Types of Calls:

- **Mimicry Calls:** Mimic specific animal sounds, such as distress calls, mating calls, or feeding calls. For example, a doe bleat can attract bucks during the rut, and a predator distress call can lure in predators.

- **Electronics Calls:** Electronic callers use recorded sounds to mimic various animal noises. They offer the advantage of volume control and a range of sounds, such as prey distress calls or predator calls.

- **Manual Calls:** Hand-operated calls, such as mouth calls or reed calls, require skill to produce realistic sounds. They are often portable and can be used for a variety of animals.

ii. Selecting the Right Call:

- **Species-Specific Calls:** Choose calls that are appropriate for the target species. For example, use turkey calls for turkeys, elk calls for elk, and coyote calls for coyotes.

- **Timing and Season:** Use calls that match the animal's behavior during different seasons or times of the day. For example, use mating calls during breeding seasons and distress calls when predators are active.

iii. Using Calls Effectively:

- **Volume Control:** Adjust the volume of the call based on distance and environmental conditions. Loud calls may carry farther but can also alert wary animals.

- **Realistic Sounds:** Ensure the call mimics the natural sound of the animal accurately. Practice using the call to produce realistic sounds that match the animal's vocalizations.

- **Pacing and Variation:** Vary the rhythm and pitch of the call to avoid sounding mechanical. Animals are more likely to respond to natural, varied calls.

2. Understanding Decoys and Their Uses

i. Types of Decoys:

- **Visual Decoys:** Representations of animals, such as inflatable or rigid decoys, are used to attract animals by mimicking their appearance. For example, a deer decoy can attract bucks during the rut.

- **Motion Decoys:** Incorporate movement, such as spinning wings or flapping feathers, to simulate live animals and draw attention. These are often used for waterfowl hunting.

- **Scent Decoys:** Utilize scent dispensers to mimic animal scents, such as estrus scents for deer or predator scents for attracting prey.

ii. Selecting the Right Decoy:

- **Species-Specific Decoys:** Choose decoys that match the target species and the specific situation. For example, use a hen decoy for turkeys or a buck decoy for attracting other bucks.

- **Environmental Matching:** Ensure the decoy matches the environment where it will be used. For example, use a realistic decoy that blends with the surroundings to avoid detection.

iii. Using Decoys Effectively:

- **Placement:** Position the decoy where it will be most visible to the target animal. Place it in an open area or near a feeding or mating site to maximize its effectiveness.

- **Movement:** Use motion decoys in areas where movement is likely to attract attention. For example, a spinning wing decoy can simulate a feeding bird and draw waterfowl closer.

- **Combination with Calls:** Use decoys in conjunction with calls to create a more convincing setup. For example, combine a turkey decoy with a turkey call to simulate a mating pair.

3. Strategic Placement and Setup

i. Location:

- **High-Traffic Areas:** Place calls and decoys in areas where animals are known to travel or gather, such as trails, feeding areas, or mating sites.

- **Concealment:** Ensure that calls and decoys are visible but not easily detected by the target animal. Use natural cover or camouflage to hide your setup from the animal's view.

ii. Wind and Visibility:

- **Wind Direction:** Position decoys and calls considering the wind direction to avoid alerting animals with your scent. Place the call or decoy upwind from where you expect the animal to approach.

- **Visibility:** Ensure the decoy is positioned where it will be visible to the target animal from a distance. Avoid placing it in areas where it may be obstructed by vegetation or terrain.

iii. Avoiding Detection:

- **Camouflage:** Use camouflage patterns or materials to blend the call and decoy with the environment. This reduces the chance of the animal detecting unnatural shapes or colors.

- **Movement Control:** Minimize any unnecessary movement that could alert animals. Set up decoys and calls before the hunt and avoid adjusting them while animals are in the vicinity.

4. Maintenance and Care

i. Decoy Maintenance:

- **Regular Cleaning:** Keep decoys clean and free from dirt or debris that could affect their appearance or function. Regularly check for damage and repair or replace decoys as needed.

- **Proper Storage:** Store decoys in a dry, cool place to prevent damage and prolong their lifespan. Avoid exposing them to harsh weather conditions or sunlight.

ii. Call Maintenance:

- **Cleaning:** Clean manual calls regularly to maintain their effectiveness and prevent clogging. Follow the manufacturer's instructions for cleaning and maintenance.

- **Battery Care:** For electronic calls, ensure that batteries are fresh and properly installed. Replace batteries before heading out to avoid malfunctions.

Using calls and decoys effectively involves understanding their types, selecting the right ones for your target species, and employing strategic placement and setup. By combining these tools with knowledge of animal behavior and ethical considerations, hunters can enhance their success while ensuring humane and responsible practices. Regular maintenance and care of calls and decoys help ensure their effectiveness and longevity.

4.3 Best Practices for Ethical Hunting

Ethical hunting involves respecting wildlife, following regulations, and ensuring humane practices throughout the hunting process. Adhering to ethical principles helps maintain ecological balance, supports conservation efforts, and fosters respect for the natural world. Here's a comprehensive guide to best practices for ethical hunting:

1. Understanding and Adhering to Regulations

i. Legal Compliance:

- **Licensing and Permits:** Obtain all necessary licenses and permits before hunting. Ensure you are aware of the specific regulations for the species you are hunting, including season dates, bag limits, and restricted areas.

- **Local Laws:** Familiarize yourself with local hunting laws and regulations, which can vary by state, region, or country. Follow all legal requirements to ensure compliance and contribute to wildlife conservation efforts.

ii. Species-Specific Regulations:

- **Protected Species:** Avoid hunting species that are protected by law or listed as endangered. Respect conservation efforts and contribute to the protection of vulnerable wildlife.

- **Harvest Limits:** Adhere to harvest limits and quotas set by wildlife management authorities. These limits help manage populations and ensure sustainable hunting practices.

2. Ensuring Humane Harvesting

i. Shot Placement:

- **Vital Areas:** Aim for vital areas such as the heart or lungs to ensure a quick and humane kill. Understanding animal anatomy helps in making ethical shots that minimize suffering.

- **Distance and Accuracy:** Ensure you are close enough to make an accurate shot. Avoid taking shots that are beyond your effective range or that you are not confident in making.

ii. Field Dressing:

- **Immediate Action:** Field dress the animal as soon as possible to prevent spoilage and ensure a more humane process. Proper field dressing also helps in maintaining meat quality.

- **Clean Procedures:** Use clean and humane methods for dressing and processing the animal. Avoid unnecessary handling or stress that could affect the meat.

3. Respecting Wildlife and Habitat

i. Respect for Animals:

- **Avoid Unnecessary Suffering:** Aim to minimize the suffering of the animal by making quick, clean kills. Avoid prolonged tracking or chasing that could cause undue stress.

- **Non-Target Species:** Be cautious not to accidentally hunt non-target or protected species. Ensure that you correctly identify your target before taking a shot.

ii. Habitat Preservation:

- **Minimize Impact:** Avoid damaging the environment or disturbing other wildlife. Follow Leave No Trace principles by cleaning up after yourself and minimizing your impact on the habitat.
- **Respect Boundaries:** Do not trespass on private property without permission. Respect property boundaries and adhere to any specific rules or requests from landowners.

4. Using the Entire Animal

i. Utilization of Resources:

- **Meat Processing:** Ensure that the meat from your harvest is fully utilized. Process and consume as much of the animal as possible to honor the life taken and reduce waste.
- **Other Parts:** Consider using other parts of the animal, such as hides, bones, and antlers, for various purposes. This approach maximizes the use of the harvested animal and supports ethical practices.

ii. Donation Programs:

- **Food Banks:** Many regions have programs that accept donated game meat for food banks or charitable organizations. Consider donating excess meat to help those in need and support community efforts.

5. Safety and Preparedness

i. Hunter Safety:

- **Safety Training:** Complete hunter safety courses to ensure you are knowledgeable about safe hunting practices and firearm handling. This training helps prevent accidents and ensures safe interactions with others in the field.

- **Safety Gear:** Wear appropriate safety gear, such as blaze orange clothing, to increase visibility and reduce the risk of accidents. Ensure that your gear is in good condition and suitable for the hunting environment.

ii. Preparation:

- **Equipment Check:** Regularly inspect and maintain your hunting equipment to ensure it is in proper working order. This includes firearms, bows, and hunting gear.

- **Emergency Plan:** Have a plan for emergencies, including first aid, navigation, and communication. Carry a first aid kit, map, compass, or GPS device, and inform someone of your hunting plans and location.

6. Ethical Conduct in the Field

i. Respect for Others:

- **Public Land Etiquette:** Share public hunting areas respectfully with other hunters and outdoor enthusiasts. Follow common courtesy and adhere to established hunting practices.

- **Handling Disputes:** Address any disputes or conflicts with other hunters calmly and respectfully. Ensure that interactions are handled in a manner that upholds ethical standards and promotes a positive hunting experience.

ii. Education and Advocacy:

- **Promote Ethics:** Advocate for ethical hunting practices within the hunting community. Share knowledge and encourage others to follow ethical guidelines to support responsible hunting.

- **Continuous Learning:** Stay informed about new developments in wildlife management, hunting regulations, and ethical practices. Engage in ongoing education to improve your skills and understanding of ethical hunting.

7. Conservation and Stewardship

i. Conservation Efforts:

- **Support Programs:** Support wildlife conservation programs and organizations that work to protect habitats, manage wildlife populations, and promote sustainable hunting practices.

- **Habitat Restoration:** Participate in or support habitat restoration efforts to enhance wildlife habitats and contribute to the overall health of ecosystems.

ii. Sustainable Practices:

- **Population Management:** Participate in hunting programs that contribute to wildlife population management and conservation goals. Understand how your hunting activities fit into broader conservation strategies.

- **Ethical Consumption:** Be mindful of the environmental impact of your hunting practices and strive to minimize any negative effects on wildlife and habitats.

Best practices for ethical hunting involve adhering to regulations, ensuring humane harvesting, respecting wildlife and habitats, and utilizing resources responsibly. By prioritizing safety, preparation, and ethical conduct, hunters contribute to conservation efforts and foster a respectful relationship with the natural world. Ongoing education, advocacy, and support for conservation programs further enhance the ethical dimensions of hunting and promote sustainable practices.

CHAPTER FIVE

Handling and Processing Caught Animals

5.1 Initial Handling and Field Dressing

Initial handling and field dressing are crucial steps immediately following the harvest of an animal. These processes ensure that the meat is preserved in optimal condition and that the animal is treated with respect. Proper handling and dressing also help to maintain hygiene, prevent spoilage, and facilitate the subsequent processing of the meat. Here's an extensive guide on initial handling and field dressing:

1. Initial Handling

i. Immediate Action:

- **Minimize Stress:** Once the animal is harvested, approach it calmly and handle it gently to minimize further stress. Ensure that you are prepared to start field dressing as soon as possible to prevent meat spoilage.

- **Assess the Situation:** Quickly assess the condition of the animal, including the location of the harvest, the level of blood loss, and any potential damage to the meat. This assessment helps in planning the field dressing process.

ii. Safety Precautions:

- **Personal Safety:** Wear gloves to protect yourself from blood and potential pathogens. Ensure that you use a clean, sharp knife and other necessary tools for field dressing.

- **Clean Environment:** If possible, work in a clean and sanitary area to avoid contamination. Avoid placing the animal directly on the ground if it is muddy or dirty.

iii. Preserving Meat Quality:

- **Cool the Animal:** Begin cooling the animal as soon as possible to slow the growth of bacteria. If the weather is warm, consider removing the internal organs and allowing the body cavity to cool down quickly.

- **Avoid Contamination:** Ensure that the carcass does not come into contact with dirt, vegetation, or other contaminants. Keep the meat clean throughout the handling and dressing process.

2. Field Dressing

i. Preparation:

- **Positioning the Animal:** Position the animal on its back or side to facilitate access to the abdominal cavity. If working alone, use a rope or strap to help position the animal securely.

- **Tools and Equipment:** Ensure you have the necessary tools for field dressing, including a sharp knife, a bone saw (if needed), and a field dressing kit. Make sure all tools are clean and in good condition.

ii. Cutting Procedures:

- **Making the Initial Cut:** Start by making a shallow incision along the midline of the abdomen from the rib cage to the pelvis. Be careful not to cut too deeply to avoid puncturing the stomach or intestines.

- **Removing the Internal Organs:** Gently separate and remove the internal organs, starting with the stomach and intestines. Use your knife carefully to cut through any connective tissues or ligaments. Avoid spilling the contents of the stomach or intestines onto the meat.

- **Removing the Heart and Lungs:** If necessary, remove the heart and lungs by cutting through the connecting tissues. Ensure that these organs are handled carefully to avoid contamination.

ii. Cleaning the Cavity:

- **Removing Residual Material:** Use a clean cloth or paper towel to wipe out any residual blood or tissue from the body cavity. This helps to prevent spoilage and maintain meat quality.

- **Rinsing (if applicable):** In some cases, you may choose to rinse the cavity with clean water. However, be cautious with rinsing, as excessive water can lead to meat spoilage. Pat the cavity dry with a clean cloth if you use water.

iii. Cooling the Carcass:

- **Airflow:** Allow adequate airflow around the carcass to promote cooling. If the weather is hot, consider hanging the carcass in a shaded area or using a portable game cooler to speed up the cooling process.

- **Avoid Direct Sunlight:** Keep the carcass out of direct sunlight to prevent rapid warming and bacterial growth.

3. Handling Special Cases

i. Dealing with Wounds:

- **Inspect for Damage:** Inspect the carcass for any damage caused by the shot or injury. If there are significant wounds, handle the affected areas carefully to avoid contamination.

- **Trimming Damaged Areas:** Trim away any damaged or bloodshot meat to ensure the quality of the final product. Discard any contaminated portions.

ii. Handling Larger Game:

- **Assistance:** For larger game, such as elk or moose, consider enlisting help to manage the field dressing process. Proper handling and lifting techniques are essential to prevent injury and ensure an efficient dressing process.

- **Additional Tools:** Larger game may require additional tools, such as a bone saw, to assist with the removal of large bones or joints.

4. Respect for the Animal

i. Ethical Considerations:

- **Respectful Handling:** Treat the animal with respect throughout the field dressing process. Handle the carcass gently and avoid unnecessary roughness.

- **Proper Disposal:** Dispose of internal organs and other waste materials properly. Follow local regulations for the disposal of animal remains and avoid leaving waste in the field.

ii. Using the Entire Animal:

- **Utilize All Parts:** Consider using all edible parts of the animal, including meat, organs, and bones. This approach maximizes the use of the animal and aligns with ethical hunting practices.

5. Post-Field Dressing

i. Transporting the Carcass:

- **Secure Transport:** Ensure that the carcass is securely transported to your processing area. Use a game bag or other protective covering to prevent contamination during transportation.

- **Processing:** Once at the processing location, follow proper procedures for butchering, aging, and preserving the meat. Maintain cleanliness and hygiene throughout the processing process.

ii. Record Keeping:

- **Documentation:** Keep records of your harvest, including date, location, and any relevant details. This information can be valuable for tracking your hunting activities and complying with regulations.

Initial handling and field dressing are essential steps in ensuring the quality and safety of harvested game meat. By following proper procedures for handling, dressing, and cooling, hunters can preserve meat quality, respect the animal, and minimize spoilage. Ethical practices, careful handling, and respect for the animal contribute to a responsible and successful hunting experience.

5.2 Transporting Game Safely

Transporting game safely is crucial for maintaining the quality of the meat, ensuring compliance with regulations, and respecting ethical considerations. Proper transportation practices prevent contamination, spoilage, and ensure that the game is handled in a manner that reflects respect for the animal. Here's a comprehensive guide on how to transport game safely:

1. Preparation for Transport

i. Game Bagging:

- **Game Bags:** Use game bags or similar protective covers to encase the carcass. Game bags are designed to protect the meat from contamination while allowing air circulation to help with cooling.

- **Clean Bags:** Ensure that the game bags are clean and free from dirt or debris. Avoid using plastic bags, as they can trap moisture and cause the meat to spoil.

ii. Cooling the Carcass:

- **Pre-Transport Cooling:** Ensure that the carcass is well-cooled before transportation. If possible, hang the carcass in a cool, shaded area to allow it to reach a lower temperature.

- **Avoid Heat:** Prevent exposure to heat during transportation. High temperatures can accelerate bacterial growth and spoilage.

2. Transporting in a Vehicle

i. Vehicle Preparation:

- **Clean Interior:** Ensure that the interior of the vehicle is clean before transporting the game. Remove any items that could potentially contaminate the carcass.

- **Ventilation:** Ensure adequate ventilation in the vehicle to help with cooling. Avoid placing the carcass in a confined space where heat could accumulate.

ii. Securing the Carcass:

- **Secure Placement:** Place the carcass in the vehicle in a manner that prevents movement during transport. Use straps or tie-downs to secure the carcass and prevent it from shifting.

- **Prevent Contamination:** Avoid placing the carcass directly on vehicle seats or surfaces that are not clean. Use a tarp or other protective layer if necessary.

Temperature Control:

- **Coolers:** Use coolers or portable refrigeration units to maintain a low temperature during transportation, especially in warm weather. Place ice packs or ice inside the cooler to help keep the carcass cool.

- **Monitor Temperature:** If transporting for an extended period, monitor the temperature of the cooler or vehicle to ensure that it remains within a safe range for preserving meat quality.

3. Transporting Large Game

i. Handling Larger Animals:

- **Additional Assistance:** For large game such as elk or moose, consider using additional assistance or specialized equipment. Lifting devices or winches can help with maneuvering the carcass into the vehicle.

- **Dismembering:** If the carcass is too large to fit in the vehicle, consider dismembering it into manageable sections. This can make transportation easier and help with cooling.

ii. Specialized Vehicles:

- **ATVs and UTVs:** For remote areas, ATVs or UTVs with cargo racks can be useful for transporting game. Ensure that the cargo area is clean and that the carcass is secured properly.

- **Trailers:** Use trailers equipped for game transport if available. Ensure that the trailer is clean and has proper ventilation.

4. Compliance with Regulations

i. Legal Requirements:

- **Check Regulations:** Be aware of and comply with local regulations regarding game transportation. Some areas may have specific rules about how game must be transported or documented.

- **Transport Permits:** In some cases, transport permits or tags may be required. Ensure that you have all necessary documentation and permits for transporting game.

ii. Interstate and International Transport:

- **Check for Restrictions:** If transporting game across state or national borders, check for specific restrictions or requirements. Different regions may have regulations regarding the import and export of game meat.

5. Ethical Considerations

i. Respect for the Animal:

- **Minimize Stress:** Handle the carcass gently to avoid causing additional stress or damage. Ensure that transportation methods are as humane as possible.

- **Dispose of Waste Properly:** Dispose of any waste materials, such as internal organs or other byproducts, in accordance with local regulations. Avoid leaving waste in public areas or natural environments.

ii. Efficient Use:

- **Utilize All Parts:** Aim to utilize as much of the animal as possible. Properly transport and process all edible parts, including meat, organs, and bones.

6. Post-Transportation Processing

i. Immediate Processing:

- **Processing Location:** Upon reaching your processing location, promptly begin butchering and processing the meat. Maintain cleanliness and hygiene throughout the processing.

- **Temperature Control:** Continue to monitor and maintain proper temperature control during processing to ensure the quality and safety of the meat.

Safe and effective transportation of game involves careful preparation, securing the carcass, and maintaining temperature control. By following proper procedures, ensuring compliance with regulations, and addressing ethical considerations, hunters can preserve the quality of the meat and handle the game respectfully. Proper handling during transportation supports both the quality of the harvest and the principles of ethical hunting.

5.3 Preparing for Skinning and Butchering

Preparing for skinning and butchering is a critical stage in handling game, ensuring that the meat is processed efficiently and remains in optimal condition. Proper preparation helps maintain hygiene, prevent contamination, and facilitate a smooth and effective processing process. Here's a detailed guide on preparing for skinning and butchering:

1. Workspace Preparation

i. Selecting a Clean Area:

- **Clean Surface:** Choose a clean, flat surface for processing the game. A dedicated processing table or clean ground area covered with a clean tarp or cloth is ideal.

- **Sanitary Conditions:** Ensure the workspace is free from dirt, debris, and other contaminants. Clean and sanitize all surfaces and tools before starting.

ii. Gathering Tools and Equipment:

- **Knives:** Use a sharp, sturdy knife for skinning and a separate boning knife for butchering. Ensure that knives are clean and properly sharpened.

- **Bone Saw:** A bone saw or reciprocating saw may be needed for larger game to cut through bones. Ensure it is in good working condition.

- **Skinning Tools:** Consider using skinning pliers or hook tools to assist with removing the hide.
- **Cutting Boards:** Have cutting boards available for butchering the meat. Use separate boards for different types of meat (e.g., one for muscle meat, another for organs).

iii. Protective Gear:

- **Gloves:** Wear disposable or durable gloves to protect your hands from blood, bacteria, and contaminants.
- **Apron:** Use an apron or protective clothing to keep your clothes clean and prevent contamination.

2. Preparing the Carcass

i. Inspecting the Carcass:

- **Check for Contamination:** Inspect the carcass for any signs of contamination or damage. Trim away any visibly spoiled or damaged areas.
- **Cool the Carcass:** Ensure that the carcass is well-cooled before starting the skinning and butchering process. A cold carcass is easier to handle and less likely to spoil.

ii. Positioning the Carcass:

- **Stable Position:** Position the carcass securely to prevent it from moving during skinning and butchering. Use a game hanger, processing table, or a similar setup to keep the carcass stable.

- **Accessibility:** Ensure that the carcass is accessible from all sides for efficient skinning and butchering. Adjust the position as needed to allow easy access.

3. Skinning Preparation

i. Tools and Techniques:

- **Initial Cut:** Begin by making an initial cut along the legs and belly to separate the hide from the meat. Use a sharp knife to avoid tearing the hide.

- **Removing the Hide:** Carefully peel the hide away from the carcass, starting from the initial cut and working your way around. Use skinning pliers or hook tools to help with pulling the hide if necessary.

- **Handling:** Handle the hide gently to avoid damaging the meat. Keep the hide clean and avoid contamination from the ground or other surfaces.

ii. Preserving the Hide:

- **Clean and Dry:** If planning to use the hide, keep it clean and dry. Avoid leaving it in direct sunlight or allowing it to become wet.

- **Salt and Freeze:** If not processing the hide immediately, apply salt to preserve it or freeze it to prevent spoilage.

4. Butchering Preparation

i. Cutting and Separation:

- **Legs and Quartering:** Begin by separating the legs and quarters from the main body. Use a bone saw or a heavy knife to cut through joints and bones.

- **Removing Organs:** Carefully remove the organs from the body cavity. Separate them from the meat and place them in a clean container if they are to be used or processed.

- **Trimming:** Trim away any excess fat, connective tissue, or sinew from the meat. This helps improve the quality and appearance of the final cuts.

ii. Organizing Cuts:

- **Meat Cuts:** Plan the cuts based on your intended use of the meat. Common cuts include steaks, roasts, and ground meat. Organize and label different cuts for easy identification and processing.

- **Bone Handling:** Separate bones from meat cuts as needed. Consider saving bones for making stock or other uses.

5. Maintaining Hygiene

i. Sanitation Practices:

- **Clean Tools:** Clean and sanitize all tools and equipment frequently during the process. Use hot, soapy water and disinfectant solutions as needed.

- **Avoid Cross-Contamination:** Use separate tools and surfaces for different types of meat (e.g., muscle meat, organs). Avoid cross-contamination by washing hands and equipment thoroughly between tasks.

ii. Safe Handling:

- **Minimize Contact:** Minimize direct contact between meat and other surfaces. Use clean utensils and avoid touching the meat with bare hands as much as possible.

- **Proper Storage:** Store processed meat in clean, airtight containers or wrap it properly to prevent contamination and spoilage.

6. Post-Processing

i. Cooling and Aging:

- **Cool Quickly:** After butchering, continue to cool the meat rapidly to prevent bacterial growth. Use refrigeration or a cool environment to maintain meat quality.

- **Aging:** If desired, age the meat in a controlled environment to enhance tenderness and flavor. Follow recommended aging times and conditions for the specific type of meat.

ii. Packaging and Freezing:

- **Proper Packaging:** Package meat in vacuum-sealed bags, freezer paper, or other appropriate packaging materials. Label packages with the date and cut type for easy identification.

- **Freeze Immediately:** Place packaged meat in the freezer as soon as possible to preserve quality and prevent freezer burn.

Preparing for skinning and butchering involves careful workspace setup, appropriate tool selection, and maintaining hygiene. Proper handling and preparation of the carcass ensure efficient processing and high-quality meat. By following these steps, hunters can ensure that the meat is handled safely, respectfully, and in accordance with best practices for meat processing.

CHAPTER SIX

Skinning Techniques

6.1 Skinning Tools and Their Uses

Skinning tools are essential for efficiently and effectively removing the hide from a harvested animal. Each tool serves a specific purpose and is designed to handle different aspects of the skinning process. Here's a detailed guide to various skinning tools and their uses:

1. Skinning Knife

Description:

- **Blade Shape:** A skinning knife typically has a curved, sharp blade designed to follow the contours of the animal's body. The curve helps in peeling the hide away from the flesh with precision.
- **Handle:** The handle is usually ergonomically designed for a comfortable grip, which is crucial for control and reducing hand fatigue.

Uses:

- **Initial Cuts:** Ideal for making initial cuts around the legs, belly, and neck to start the skinning process.

- **Peeling the Hide:** Used to carefully separate the hide from the meat by sliding the blade between the skin and muscle tissue. The curved blade helps in making smooth, controlled cuts.

Best Practices:

- **Keep Sharp:** Regularly sharpen the blade to maintain efficiency and reduce the risk of tearing the hide.
- **Clean Blade:** Clean the knife thoroughly after each use to prevent contamination and ensure hygiene.

2. Boning Knife

Description:

- **Blade Shape:** A boning knife typically has a narrow, straight blade or a slight curve. It's designed to maneuver around bones and remove meat from the bone with precision.
- **Handle:** Features a handle that provides a secure grip, allowing for detailed work around joints and bones.

Uses:

- **Deboning:** Used for removing meat from bones after the hide has been removed. It's effective for separating muscle tissue from bone and for trimming excess fat or connective tissue.
- **Detail Work:** Ideal for more delicate tasks where precision is needed, such as removing small bones or connective tissue.

Best Practices:

- **Sharp Blade:** Ensure the blade is sharp to avoid damaging the meat and to facilitate precise cuts.

- **Hygiene:** Clean and sanitize the boning knife between tasks to prevent cross-contamination.

3. Skinning Pliers

Description:

- **Design:** Skinning pliers are designed with a grip handle and serrated or textured jaws to help pull the hide away from the meat.

- **Material:** Made from durable metal, they are built to handle the tension and pressure involved in pulling the hide.

Uses:

- **Peeling the Hide:** Useful for gripping and pulling the hide away from the carcass, especially in areas where the hide is stubborn or tightly adhered.

- **Assisting in Skinning:** Helps in maintaining a steady grip on the hide while cutting around it.

Best Practices:

- **Proper Use:** Use skinning pliers in conjunction with a skinning knife for best results. Avoid using excessive force that might tear the hide.

- **Maintenance:** Keep the pliers clean and inspect them regularly for damage or wear.

4. Bone Saw

Description:

- **Type:** A bone saw is a tool designed for cutting through bone and larger joints. It can be a hand-operated saw or a powered saw, depending on the size of the game.
- **Blade:** The blade is serrated to handle tough, dense bone material.

Uses:

- **Cutting Joints:** Ideal for cutting through large bones or joints when disassembling the carcass, especially for larger game like deer, elk, or moose.
- **Sectioning Carcass:** Useful for dividing the carcass into more manageable sections for easier handling and processing.

Best Practices:

- **Safety:** Use the bone saw carefully to avoid injury and to ensure clean cuts. Ensure the saw is properly maintained and sharpened.

- **Clean After Use:** Clean the saw thoroughly after each use to prevent the spread of bacteria and to maintain its effectiveness.

5. Fleshing Knife

Description:

- **Blade Shape:** A fleshing knife has a slightly curved blade that's used to scrape away remaining flesh, fat, and connective tissue from the hide after it has been removed.
- **Handle:** Designed for comfortable use while scraping the hide.

Uses:

- **Cleaning the Hide:** Used to remove any remaining muscle tissue or fat from the inside of the hide to prepare it for preservation or tanning.
- **Final Touches:** Helps in preparing the hide for further processing by ensuring it is clean and free of excess flesh.

Best Practices:

- **Sharpness:** Keep the fleshing knife sharp to make the scraping process easier and more effective.
- **Gentle Pressure:** Use gentle pressure to avoid damaging the hide while removing excess tissue.

6. Game Shears

Description:

- **Design:** Game shears are heavy-duty scissors designed specifically for cutting through smaller bones, cartilage, and tendons.
- **Material:** Typically made of stainless steel or high-strength metal for durability.

Uses:

- **Cutting Cartilage:** Useful for cutting through cartilage and tendons that are difficult to cut with a knife.
- **Small Joints:** Ideal for separating smaller joints or sections of the carcass.

Best Practices:

- **Keep Clean:** Clean the shears after each use to prevent contamination and maintain sharpness.
- **Safety:** Use game shears with caution to avoid accidental cuts or injuries.

7. Cutting Board

Description:

- **Material:** Cutting boards are usually made from wood, plastic, or composite materials. They provide a clean surface for cutting and processing meat.
- **Size:** Select a board large enough to handle the size of the game you are processing.

Uses:

- **Meat Processing:** Provides a sanitary surface for butchering and processing meat. Helps in organizing cuts and minimizing contamination.

- **Separation of Cuts:** Use separate boards for different types of meat (e.g., muscle meat, organs) to avoid cross-contamination.

Best Practices:

- **Sanitize Regularly:** Clean and sanitize the cutting board between uses. Use separate boards for different types of meat to prevent cross-contamination.

- **Stable Surface:** Ensure the cutting board is stable and doesn't slide during use. Consider using a non-slip mat underneath.

Each skinning tool plays a specific role in the process of removing the hide and preparing the meat for further processing. By selecting the appropriate tools and using them correctly, hunters can ensure an efficient and effective skinning process, maintain meat quality, and adhere to best practices for hygiene and safety.

6.2 Step-by-Step Skinning Process

Skinning a harvested animal requires a methodical approach to ensure the hide is removed cleanly while preserving the quality of the meat. Here's a detailed, step-by-step guide to the skinning process:

1. Preparation

i. Workspace Setup:

- **Clean Area:** Ensure your work area is clean, flat, and free from contaminants. Use a clean tarp or processing table to work on.
- **Tools Ready:** Gather all necessary tools, including a skinning knife, bone saw, skinning pliers, and a cutting board. Ensure they are sharp and clean.

ii. Personal Gear:

- **Protective Clothing:** Wear disposable gloves and an apron to protect yourself and maintain hygiene.
- **Safety First:** Ensure the knife is sharp and handle it with care to prevent injury.

2. Initial Cuts

i. Positioning the Carcass:

- **Stabilize the Carcass:** Hang the carcass by the hind legs or place it on a processing table. Ensure it is stable and accessible from all sides.

ii. Making Initial Cuts:

- **Around the Legs:** Start by making cuts around the hind legs and shoulders, connecting these cuts down the belly. Be careful not to cut too deep; aim to just penetrate the skin.

- **Along the Belly:** Cut along the belly from the sternum to the pelvis, following the contour of the body. Avoid cutting into the intestines or other internal organs.

3. Peeling the Hide

i. Starting the Peel:

- **Loosen the Hide:** Use the skinning knife to gently loosen the hide from the meat along the initial cuts. Slide the knife between the hide and the flesh, taking care to separate them without tearing.

- **Use Pliers:** For stubborn areas, use skinning pliers to grip and pull the hide away from the carcass while cutting.

ii. Working Around the Carcass:

- **Peel from Top Down:** Continue peeling the hide away from the body, working from the top (back) downward. Work carefully around the legs and shoulders, using the knife to separate the hide from the meat.

- **Avoid Tearing:** Keep the hide as intact as possible. Use gentle, consistent pressure to avoid tearing, which can make removal more difficult and affect the quality of the hide.

4. Removing the Hide

i. Legs and Shoulders:

- **Cut Around Joints:** Use the knife to cut around the joints at the shoulders and hind legs. This allows you to free the hide from these areas more easily.

- **Work Slowly:** Continue working around these joints, carefully separating the hide from the meat and bone.

ii. Final Steps:

- **Remove Hide:** Once the hide is mostly peeled off, carefully pull it away from the carcass. Use skinning pliers or your hands to assist in pulling it off without causing damage.

- **Check for Stubborn Areas:** If any parts of the hide are still attached, use the knife to carefully separate them.

5. Handling the Hide

i. Inspection:

- **Check for Damage:** Inspect the hide for any tears or damage. If there are significant tears, you may need to trim these areas or address them during the preservation process.

- **Clean the Hide:** If the hide is to be preserved, clean it by removing any blood or debris. Avoid soaking or washing the hide excessively, as this can affect its quality.

ii. Preservation:

- **Salt the Hide:** If you plan to preserve the hide, apply a generous amount of salt to it to prevent bacterial growth and spoilage. Ensure that the entire surface is covered.
- **Freeze or Dry:** Alternatively, you can freeze the hide or dry it as per your preservation method.

6. Post-Skinning Care

i. Inspect the Carcass:

- **Check for Contamination:** Once the hide is removed, inspect the carcass for any signs of contamination or spoilage. Trim away any affected areas.
- **Cool the Meat:** Continue to cool the carcass rapidly if it will not be processed immediately. Maintain proper hygiene throughout.

ii. Clean Up:

- **Sanitize Tools:** Clean and sanitize all tools and surfaces used during the skinning process. Proper cleaning helps prevent contamination and prepares the tools for future use.
- **Dispose of Waste:** Properly dispose of any waste materials, including the hide if not preserving it. Follow local regulations for waste disposal.

The skinning process involves careful preparation, making initial cuts, peeling the hide, and handling it properly for preservation. By following these steps, hunters can ensure a clean, efficient skinning process that maintains the quality of both the meat and the hide. Proper technique and hygiene are essential to achieving the best results and respecting the harvested animal.

6.3 Handling Different Types of Fur and Hide

Handling various types of fur and hide requires specific techniques to ensure their preservation and quality. Different animals have different fur and hide characteristics, and understanding these can help in processing, preserving, and utilizing them effectively. Here's a comprehensive guide to handling different types of fur and hide:

1. Deer Hide

Characteristics:

- **Texture:** Deer hide is generally soft and pliable, with a relatively thin layer of fur.
- **Fur:** The fur is short and dense, providing good insulation but requiring careful handling to avoid damage.

Handling Tips:

- **Skinning:** Handle the skin carefully to avoid tearing. Deer hide is more delicate than that of larger game, so use a sharp, curved skinning knife.
- **Preservation:** For tanning, salt the hide thoroughly to prevent bacterial growth. You can also use a commercial tanning solution or freeze it if you plan to tan it later.

- **Cleaning:** Clean the hide with a damp cloth or sponge to remove any blood or debris. Avoid excessive washing.

2. Elk Hide

Characteristics:

- **Texture:** Elk hide is thicker and more robust than deer hide, with a coarser texture.
- **Fur:** The fur is longer and denser, providing excellent insulation but requiring more effort to handle.

Handling Tips:

- **Skinning:** Use a sturdy, sharp knife and a bone saw for larger joints. Elk hide requires more effort due to its thickness.
- **Preservation:** Salt the hide generously and consider using a commercial tanning product for best results. Alternatively, freeze it if not processing immediately.
- **Cleaning:** Use a soft brush or cloth to clean off blood and debris. Avoid soaking the hide.

3. Bear Hide

Characteristics:

- **Texture:** Bear hide is thick and tough, with dense fur that can be challenging to handle.

- **Fur:** The fur is long and coarse, requiring special attention during skinning and handling.

Handling Tips:

- **Skinning:** Employ a heavy-duty skinning knife and possibly a skinning tool to manage the tough hide. Pay special attention to avoid tearing the hide.

- **Preservation:** Salt the hide thoroughly and use a tanning solution appropriate for bear hides. Bears often have a higher fat content, so thorough salting is crucial.

- **Cleaning:** Clean the hide with a damp cloth and remove excess fat, as it can spoil the hide if not properly addressed.

4. Raccoon Hide

Characteristics:

- **Texture:** Raccoon hide is relatively thin but has dense fur, with a distinctive pattern and color.

- **Fur:** The fur is soft and can be prone to damage if not handled carefully.

Handling Tips:

- **Skinning:** Use a sharp knife to carefully separate the hide from the carcass. Raccoon hides are smaller and more delicate, so handle with care.

- **Preservation:** Salt the hide well to prevent spoilage. Raccoon hides can also be tanned using commercial products or preserved through freezing.

- **Cleaning:** Gently clean the hide with a cloth, avoiding excessive moisture that can damage the fur.

5. Fox Hide

Characteristics:

- **Texture:** Fox hide is relatively thin with fine fur that requires delicate handling.

- **Fur:** The fur is soft and plush, often used in fur garments.

Handling Tips:

- **Skinning:** Handle the hide gently and use a sharp, curved knife to avoid tearing the delicate fur.

- **Preservation:** Salt the hide thoroughly and use a specialized tanning solution for fine fur. Freezing is also an option if you plan to process later.

- **Cleaning:** Use a soft brush or cloth to clean the hide, avoiding water if possible to prevent fur damage.

6. Coyote Hide

Characteristics:

- **Texture:** Coyote hide is thick with a dense undercoat and longer guard hairs.

- **Fur:** The fur is somewhat coarse, and the hide requires careful handling to maintain its quality.

Handling Tips:

- **Skinning:** Use a sturdy knife and handle the hide carefully to avoid damage. Coyotes have a robust fur that needs proper attention.

- **Preservation:** Salt the hide thoroughly and use a commercial tanning product if desired. Freezing is a good alternative if you plan to process the hide later.

- **Cleaning:** Clean the hide with a damp cloth or sponge and avoid soaking.

7. Small Game (e.g., Rabbit, Squirrel)

Characteristics:

- **Texture:** Small game hides are thin and delicate, with soft fur.

- **Fur:** The fur is usually fine and can be easily damaged if not handled properly.

Handling Tips:

- **Skinning:** Use a small, sharp knife to make precise cuts. The hides are delicate, so work slowly and carefully.

- **Preservation:** Salt the hide immediately to prevent spoilage. Small game hides can also be frozen or tanned using appropriate methods.

- **Cleaning:** Gently clean the hide with a soft cloth, avoiding excessive moisture.

Handling different types of fur and hide requires understanding their unique characteristics and applying appropriate techniques for skinning, preservation, and cleaning. By using the right tools and methods for each type of hide, hunters and processors can ensure high-quality results and make the most of their harvest. Whether dealing with thick, coarse hides or delicate, fine furs, proper handling practices are key to maintaining the integrity and usability of the hides.

CHAPTER SEVEN

Tanning and Preserving Hides

7.1 Introduction to Tanning Methods

Tanning is the process of preserving animal hides and turning them into leather. This crucial step not only prevents decay but also transforms the hide into a durable, flexible material suitable for various uses. Understanding the different tanning methods and their applications can help you choose the best method for your needs and ensure the quality of the final product.

1. Overview of Tanning

Purpose of Tanning:

- **Preservation:** Tanning prevents decomposition by stabilizing the collagen in the hide, making it resistant to bacteria and fungi.
- **Flexibility:** It transforms the hide into a flexible material that can be used for clothing, upholstery, and other products.
- **Durability:** Properly tanned hides are more durable and can withstand wear and tear better than raw hides.

Basic Tanning Process:

- **Preparation:** The hide is cleaned, salted, and soaked to remove impurities and prepare it for tanning.
- **Tanning:** The tanning agent is applied to the hide, chemically altering the collagen structure to stabilize it.
- **Finishing:** The tanned hide is dried, conditioned, and often dyed or treated to achieve the desired texture and appearance.

2. Types of Tanning Methods

1. Vegetable Tanning:

Description:

- **Agents Used:** Utilizes natural plant materials such as tannins found in tree bark, leaves, and fruits.
- **Process:** The hide is soaked in a solution containing tannins, which penetrate and react with the collagen in the hide to create a stable, flexible material.

Advantages:

- **Environmentally Friendly:** Uses natural, renewable resources and is less polluting compared to other methods.
- **Aesthetic Qualities:** Produces a leather with a natural look and feel, which can develop a rich patina over time.

Disadvantages:

- **Time-Consuming:** The process can take several weeks, making it slower compared to other methods.
- **Cost:** May be more expensive due to the time and materials involved.

Applications:

- **Traditional Leather Goods:** Used for high-quality leather items such as saddles, belts, and wallets.

2. Chrome Tanning:

Description:

- **Agents Used:** Employs chromium salts, primarily chromium sulfate, to tan the hide.
- **Process:** The hide is soaked in a solution of chromium salts, which quickly penetrate and react with the collagen, resulting in a more uniform and faster tanning process.

Advantages:

- **Speed:** The process is much faster, taking just a few days to complete.
- **Durability:** Produces leather that is resistant to water and wear, making it suitable for various applications.

Disadvantages:

- **Environmental Impact:** Chromium tanning can be more polluting and requires proper disposal and treatment of waste products.
- **Less Natural Appearance:** The leather often has a more uniform appearance and may not develop the same patina as vegetable-tanned leather.

Applications:

- **Industrial and Consumer Products:** Commonly used for automotive interiors, shoes, and fashion accessories.

3. Aldehyde Tanning:

Description:

- **Agents Used:** Utilizes aldehyde compounds such as formic or glutaraldehyde.
- **Process:** The hide is treated with aldehyde solutions that react with collagen, resulting in a softer and more pliable leather.

Advantages:

- **Softness:** Produces exceptionally soft and flexible leather.
- **Low Environmental Impact:** Generally considered to be less toxic and environmentally harmful compared to chrome tanning.

Disadvantages:

- **Durability:** May not be as durable or resistant to water as other tanning methods.
- **Cost:** Can be more expensive due to the cost of chemicals and processing.

Applications:

- **Specialty Leather Goods:** Often used for high-end fashion items, upholstery, and garments.

4. Brain Tanning:

Description:

- **Agents Used:** Uses animal brains or other natural fats and oils as the tanning agent.
- **Process:** The hide is soaked and worked with a mixture of brains and other natural substances, which slowly penetrate and tan the hide.

Advantages:

- **Traditional Method:** Provides a historical and artisanal approach to tanning.
- **Natural Qualities:** Produces a very soft, flexible, and breathable leather with a unique texture.

Disadvantages:

- **Labor-Intensive:** The process is time-consuming and requires significant manual effort.
- **Limited Use:** Not suitable for large-scale production or modern leather goods.

Applications:

- **Traditional and Indigenous Crafts:** Often used in traditional clothing, blankets, and historical reenactment items.

3. Factors Influencing the Choice of Tanning Method

i. Hide Type: Different hides may respond better to certain tanning methods. For example, thicker hides like those from large game might be best suited to vegetable or chrome tanning.

ii. End Use: The intended use of the leather will influence the choice of tanning method. For example, durable leather for upholstery may benefit from chrome tanning, while high-quality leather for artisanal products may be better suited to vegetable tanning.

iii. Environmental Considerations: Consider the environmental impact of each tanning method. Vegetable tanning is generally more eco-friendly, while chrome tanning requires careful management of waste products.

iv. Cost and Time: Evaluate the cost and time involved in each tanning method. Vegetable tanning is slower and potentially more expensive, while chrome tanning is faster but may involve higher costs for chemicals.

Understanding the different tanning methods helps in selecting the most appropriate technique based on the type of hide, the intended use, and environmental considerations. Each method has its advantages and disadvantages, affecting the final product's appearance, durability, and cost. By choosing the right tanning method, you can ensure the quality and suitability of the leather for your specific needs.

7.2 Natural vs. Chemical Tanning

Tanning is essential for preserving animal hides and converting them into leather. The choice between natural and chemical tanning methods can significantly impact the final product's characteristics, environmental footprint, and cost. Understanding the differences between these two approaches helps in selecting the most suitable method based on the desired outcomes and practical considerations.

1. Natural Tanning

Natural tanning involves the use of natural substances derived from plants or animals to tan hides. This method has been used for thousands of years and is characterized by its reliance on traditional materials and techniques.

Key Methods:

i. Vegetable Tanning:

- **Agents Used:** Tannins from plant sources such as oak bark, chestnut, and hemlock.
- **Process:** The hide is soaked in a tannin-rich solution, which slowly penetrates and reacts with the collagen in the hide, stabilizing it.

ii. Brain Tanning:

- **Agents Used:** Animal brains, fats, or oils.
- **Process:** The hide is treated with a mixture of brains and other natural fats, which gradually permeate the hide and alter its properties.

Advantages:

- **Eco-Friendliness:** Natural tanning methods, especially vegetable tanning, are generally considered environmentally friendly. They use renewable resources and have a lower environmental impact compared to chemical methods.

- **Unique Qualities:** Produces leather with distinctive natural qualities, including unique textures and patinas that develop over time.

- **Traditional Craftsmanship:** Maintains traditional craftsmanship and is often preferred for artisanal and historical leather goods.

Disadvantages:

- **Time-Consuming:** Natural tanning methods, particularly vegetable tanning, are slower and can take several weeks to complete.

- **Cost:** May be more expensive due to the time involved and the cost of natural materials.

- **Inconsistent Results:** Can lead to variability in leather quality and appearance, depending on the source of the tanning agents and the skill of the tanner.

Applications:

- **High-Quality Leather Goods:** Often used for high-end, artisanal leather products, such as saddles, belts, and luxury accessories.

- **Traditional and Historical Products:** Preferred for items that require a traditional look and feel.

2. Chemical Tanning

Chemical tanning uses synthetic or mineral substances to rapidly tan hides. This method is more modern and includes various techniques that offer different advantages in terms of speed, consistency, and durability.

Key Methods:

i. Chrome Tanning:

- **Agents Used:** Chromium salts, primarily chromium sulfate.
- **Process:** The hide is soaked in a solution of chromium salts, which quickly penetrate and react with the collagen, resulting in a uniform and durable leather.

ii. Aldehyde Tanning:

- **Agents Used:** Aldehyde compounds such as formic or glutaraldehyde.
- **Process:** The hide is treated with aldehyde solutions, producing a soft and pliable leather.

Advantages:

- **Speed:** Chemical tanning methods, especially chrome tanning, are much faster, often completing the process in just a few days.
- **Consistency:** Provides uniform results in terms of color, texture, and quality, making it suitable for large-scale production.

- **Durability:** Produces leather that is resistant to water, wear, and various environmental factors.

Disadvantages:

- **Environmental Impact:** Chemical tanning methods can have a significant environmental impact. Chromium tanning, in particular, involves handling hazardous chemicals and waste disposal concerns.

- **Less Natural Appearance:** Leather produced by chemical tanning methods may lack the unique qualities and patina associated with natural tanning.

- **Health Concerns:** Some chemical tanning agents may pose health risks to workers and require strict safety measures.

Applications:

- **Industrial and Consumer Products:** Commonly used for items requiring durability and uniformity, such as automotive interiors, shoes, and fashion accessories.

- **Mass Production:** Ideal for large-scale production due to its speed and consistency.

3. Comparative Analysis

i. Environmental Considerations:

- **Natural Tanning:** Generally more eco-friendly, especially vegetable tanning, which uses natural materials and has a lower overall environmental impact.

- **Chemical Tanning:** Can be more polluting, requiring careful management of chemical waste and proper disposal procedures to minimize environmental harm.

ii. Cost and Efficiency:

- **Natural Tanning:** Typically more labor-intensive and time-consuming, leading to higher costs and longer processing times.

- **Chemical Tanning:** More efficient and cost-effective for large-scale production, with faster processing times and more consistent results.

iii. Leather Characteristics:

- **Natural Tanning:** Produces leather with a unique appearance and feel, which can develop a distinctive patina over time. Preferred for high-quality and artisanal products.

- **Chemical Tanning:** Offers uniformity and durability, with leather that is resistant to water and wear. Suitable for mass-produced and functional items.

Choosing between natural and chemical tanning methods involves weighing factors such as environmental impact, cost, efficiency, and the desired characteristics of the leather. Natural tanning, including vegetable and brain tanning, offers eco-friendliness and unique qualities but is slower and more labor-intensive. Chemical tanning, such as chrome and aldehyde tanning, provides speed and consistency but has higher environmental and health considerations. Understanding these differences helps in making an informed choice based on specific needs and priorities.

7.3 Step-by-Step Tanning Process

Tanning is the process of converting raw hides into durable leather through various chemical or natural methods. The specific steps can vary based on the chosen tanning method, but the general process includes preparation, tanning, and finishing. Here's a comprehensive step-by-step guide for both natural (vegetable) and chemical (chrome) tanning methods.

1. Preparation of Hides

i. Initial Cleaning

- **Remove Excess Flesh:** Use a sharp knife to carefully scrape off any remaining flesh, fat, or connective tissue from the hide. This step is crucial for ensuring the tanning agents penetrate the hide effectively.

- **Soaking:** Soak the hide in water to rehydrate and loosen any remaining debris. This is typically done in a large container or vat for several hours or overnight.

ii. Dehairing (if necessary)

- **Chemical Dehairing:** For some methods, such as chrome tanning, dehairing can be done using a chemical solution to dissolve the hair follicles.

- **Manual Dehairing:** In natural tanning, hair can be removed manually using a scraper or dehairing machine.

Ensure you remove all hair to prevent contamination of the leather.

iii. Flesh Removal

- **Scraping:** After soaking, scrape off any remaining flesh or fat from the hide using a fleshing knife or machine. This ensures a clean surface for the tanning agents.

2. Tanning

i. Vegetable Tanning

- **Preparation of Tanning Solution:** Prepare a tanning solution using tannins extracted from plant materials such as oak bark, chestnut, or hemlock. This involves boiling the plant material in water to extract the tannins.

- **Soaking the Hide:** Submerge the hide in the tannin solution, ensuring it is fully immersed. The hide is left in the solution for several weeks, with periodic stirring to ensure even penetration of the tannins.

- **Rinsing:** After the initial soak, rinse the hide in clean water to remove excess tannins and impurities.

- **Additional Treatments:** The hide may be treated with additional tannin solutions or natural oils to achieve the desired properties.

ii. Chrome Tanning

- **Preparation of Tanning Solution:** Prepare a chrome tanning solution using chromium salts, such as chromium sulfate. This solution is mixed according to specific concentrations and formulations.

- **Soaking the Hide:** Submerge the hide in the chrome tanning solution. The hide typically remains in the solution for a few hours to a couple of days, depending on the desired outcome.

- **Rinsing and Neutralization:** After tanning, rinse the hide thoroughly to remove excess chromium and neutralize any remaining acidity.

3. Post-Tanning Treatment

i. Drying

- **Air Drying:** Hang the hide in a well-ventilated area to dry naturally. Ensure it is not exposed to direct sunlight or excessive heat, which can cause uneven drying or damage.

- **Drum Drying:** For a more controlled process, the hide may be dried using a drum dryer, which rotates the hide to ensure even drying.

ii. Conditioning

- **Softening:** After drying, the leather may be conditioned to restore flexibility and suppleness. Apply leather conditioners or oils that are appropriate for the type of tanning used.

- **Buffing:** Use a leather buffing machine or cloth to smooth the surface and enhance the texture.

iii. Finishing

- **Dyeing (Optional):** If desired, apply dyes or pigments to the leather to achieve specific colors or patterns. This can be done using various methods, including immersion or spraying.

- **Sealing:** Apply a protective finish or sealant to protect the leather from moisture and wear. This can include waxes, oils, or synthetic sealers.

iv. Quality Inspection

- **Check for Defects:** Inspect the finished leather for any defects, such as uneven dyeing, tears, or blemishes. Address any issues by re-treating or patching as needed.

- **Final Conditioning:** Apply a final layer of conditioner or finish to ensure the leather is well-protected and maintains its appearance.

4. Special Considerations

i. Handling and Storage

- **Proper Storage:** Store tanned hides in a cool, dry place away from direct sunlight and humidity. Proper storage helps maintain the leather's quality and prevents deterioration.

- **Handling:** Handle tanned hides carefully to avoid scratches or damage. Use clean, soft cloths and gloves when necessary.

ii. Environmental Considerations

- **Waste Management:** Properly dispose of or treat waste products, such as used tanning solutions or chemical residues, according to local regulations to minimize environmental impact.

- **Eco-Friendly Practices:** Consider using eco-friendly tanning agents and practices to reduce the environmental footprint of the tanning process.

The tanning process involves several key steps, including preparation, tanning, and finishing, which vary depending on the chosen method. Natural tanning, such as vegetable tanning, is more traditional and eco-friendly but is slower and labor-intensive. Chemical tanning, such as chrome tanning, is faster and more consistent but may have higher environmental and health considerations. By following these detailed steps and considerations, you can produce high-quality leather suitable for various applications while addressing environmental and practical factors.

CHAPTER EIGHT

Drying and Storing Tanned Hides

8.1 Methods of Drying Hides

Drying is a crucial step in the leather production process, following the tanning stage. Proper drying ensures that the hide retains its desired properties and is ready for further processing or use. Different methods of drying hides can impact the final quality, texture, and usability of the leather. Here's a comprehensive guide to the various methods of drying hides:

1. Air Drying

Air drying is a traditional and commonly used method that involves hanging the hide in a well-ventilated area to dry naturally.

Process:

- **Preparation:** After tanning and conditioning, remove excess moisture from the hide by gently squeezing or blotting with clean cloths.

- **Hanging:** Hang the hide in a cool, dry, and well-ventilated area. Use a clothesline, drying rack, or similar setup to ensure even exposure to air.

- **Monitoring:** Regularly check the hide for even drying and adjust its position if necessary. Ensure it is not exposed to direct sunlight or excessive heat, which can cause uneven drying or damage.

Advantages:

- **Simplicity:** Requires minimal equipment and is easy to implement.

- **Cost-Effective:** Does not involve additional costs for specialized drying equipment.

- **Natural Drying:** Preserves the natural qualities of the hide, maintaining its texture and appearance.

Disadvantages:

- **Time-Consuming:** Can take several days to weeks, depending on the hide's thickness and environmental conditions.

- **Risk of Contamination:** Exposure to dust, insects, or humidity can affect the quality of the hide.

Applications:

- **Artisanal Leather:** Often used for small-scale or artisanal leather production, where natural methods are preferred.

2. Drum Drying

Drum drying involves using a specialized drum dryer to rotate and dry the hide in a controlled environment.

Process:

- **Preparation:** After tanning and conditioning, the hide is placed in a large rotating drum.
- **Drying:** The drum rotates slowly, allowing the hide to be exposed to warm air. This ensures even drying and reduces the risk of warping or uneven texture.
- **Monitoring:** The drying process is monitored to maintain the correct temperature and humidity levels within the drum.

Advantages:

- **Speed:** Much faster than air drying, with drying times typically ranging from a few hours to a day.
- **Consistency:** Provides uniform drying, reducing the risk of defects or inconsistencies in the leather.

Disadvantages:

- **Cost:** Requires investment in specialized equipment and energy for operation.
- **Complexity:** Involves more complex setup and maintenance compared to air drying.

Applications:

- **Industrial Leather Production:** Commonly used in large-scale leather manufacturing where speed and consistency are essential.

3. Vacuum Drying

Vacuum drying uses reduced pressure to accelerate the drying process, removing moisture from the hide more efficiently.

Process:

- **Preparation:** Place the hide in a vacuum chamber.
- **Drying:** The chamber is evacuated to create a low-pressure environment, which lowers the boiling point of water and speeds up the drying process.
- **Monitoring:** Control the temperature and pressure within the chamber to ensure optimal drying conditions.

Advantages:

- **Speed:** Significantly faster than air drying or drum drying, with drying times often reduced to a few hours.

- **Preservation:** Helps maintain the hide's quality and structure, minimizing the risk of shrinkage or warping.

Disadvantages:

- **Cost:** Requires investment in vacuum drying equipment and energy.
- **Complexity:** Involves more complex operation and monitoring compared to traditional methods.

Applications:

- **High-Quality Leather Production:** Used in scenarios where preserving the hide's quality and reducing drying time are critical.

4. Freeze Drying

Freeze drying, or lyophilization, involves freezing the hide and then removing moisture through sublimation under a vacuum.

Process:

- **Preparation:** Freeze the hide to solidify the moisture within it.
- **Drying:** Place the frozen hide in a vacuum chamber where the pressure is lowered, allowing the frozen moisture to sublimate (turn directly from ice to vapor).

- **Completion:** The process continues until all moisture is removed, leaving behind a dry, stable hide.

Advantages:

- **Quality Preservation:** Retains the hide's original structure and texture, with minimal distortion or shrinkage.
- **Speed:** Can be faster than air drying while providing high-quality results.

Disadvantages:

- **Cost:** Expensive due to the cost of freeze-drying equipment and energy.
- **Complexity:** Requires precise control and monitoring throughout the process.

Applications:

- **Specialty Leather Products:** Ideal for high-end or specialty leather applications where quality and texture are paramount.

5. Heat Drying

Heat drying involves using a controlled heat source to speed up the drying process.

Process:

- **Preparation:** Place the hide in a drying chamber or area equipped with a heating system.
- **Drying:** Apply controlled heat to the hide, ensuring even distribution to prevent overheating or damage.
- **Monitoring:** Regularly check the hide to maintain proper temperature and humidity levels.

Advantages:

- **Speed:** Provides faster drying compared to air drying, with reduced overall processing time.
- **Control:** Allows for precise control over the drying environment, reducing the risk of defects.

Disadvantages:

- **Risk of Damage:** Excessive heat can cause damage, such as warping or burning, if not properly controlled.
- **Cost:** Requires investment in heating equipment and energy.

Applications:

- **Commercial Leather Production:** Used in commercial settings where faster drying is needed without compromising quality.

Different methods of drying hides—air drying, drum drying, vacuum drying, freeze drying, and heat drying—each offer unique advantages and challenges. Air drying is simple and cost-effective but slow, while drum and vacuum drying provide speed and consistency but require specialized equipment. Freeze drying preserves the highest quality but is expensive, and heat drying offers a balance of speed and control. Selecting the appropriate drying method depends on the scale of production, desired quality, and available resources.

8.2 Preventing Mold and Damage in Hides

Preventing mold and damage during the drying and storage of hides is crucial to ensuring the final quality of leather. Mold growth and other forms of damage can compromise the leather's appearance, durability, and usability. Here's a comprehensive guide to preventing mold and damage throughout the drying and storage process:

1. Proper Drying Techniques

i. Air Drying

- **Ventilation:** Ensure the drying area is well-ventilated to promote air circulation and prevent the accumulation of moisture, which can lead to mold growth. Use fans or ventilation systems if necessary.

- **Avoid Direct Sunlight:** While sunlight can help with drying, prolonged exposure can cause uneven drying and damage. Use indirect light or shade to protect the hide.

- **Regular Inspection:** Check the hide frequently during the drying process for signs of mold or damage. Adjust the drying conditions as needed.

ii. Drum Drying

- **Temperature Control:** Maintain appropriate temperatures within the drum dryer to prevent overheating or uneven drying. Follow manufacturer guidelines for optimal temperature settings.

- **Even Rotation:** Ensure the drum rotates evenly to prevent areas of the hide from becoming too dry or remaining damp, which can lead to mold growth.

iii. Vacuum Drying

- **Proper Vacuum Levels:** Maintain the correct vacuum levels and temperatures to ensure effective moisture removal. Follow manufacturer guidelines to avoid excessive drying or overheating.

- **Monitor Moisture Levels:** Regularly check the hide's moisture content to ensure it is drying uniformly without excessive dryness that could lead to brittleness.

iv. Freeze Drying

- **Temperature Monitoring:** Ensure consistent freezing temperatures to prevent uneven sublimation. Follow recommended procedures to maintain the quality of the hide.

- **Avoid Contamination:** Keep the hide and drying equipment clean to prevent contamination and mold growth during the freeze-drying process.

v. Heat Drying

- **Controlled Heat:** Use controlled heat sources to prevent overheating, which can cause damage or shrinkage. Follow proper guidelines for heat application.

- **Regular Checks:** Frequently inspect the hide to ensure it is drying evenly and adjust the heat settings as needed to prevent mold and damage.

2. Effective Storage Practices

i. Clean Storage Environment

- **Cleanliness:** Ensure the storage area is clean and free from dust, debris, and other contaminants that could contribute to mold growth.

- **Humidity Control:** Maintain a low-humidity environment to prevent mold and mildew. Use dehumidifiers if necessary to control moisture levels.

ii. Temperature Management

- **Cool Storage:** Store hides in a cool environment, ideally between 50°F and 70°F (10°C to 21°C). Avoid high temperatures that can cause deterioration or mold growth.

- **Stable Conditions:** Maintain stable temperature conditions to prevent fluctuations that can affect the hide's quality.

iii. Proper Handling and Packaging

- **Avoid Moisture Exposure:** Ensure hides are fully dried before storage to prevent moisture-related issues. Avoid contact with water or damp environments.

- **Protective Packaging:** Use breathable packaging materials, such as cotton or canvas, to protect the hide while allowing for air circulation. Avoid plastic, which can trap moisture and lead to mold growth.

3. Mold Prevention and Treatment

i. Preventive Measures

- **Antimicrobial Agents:** Consider using antimicrobial treatments or sprays designed for leather to prevent mold growth. Follow the product instructions for safe application.

- **Regular Inspections:** Periodically inspect stored hides for signs of mold or mildew. Early detection allows for prompt treatment and prevents widespread damage.

ii. Mold Removal

- **Brush Off Mold:** For minor mold growth, use a soft brush or cloth to gently remove surface mold. Avoid scrubbing, which can damage the hide.

- **Cleaning Solution:** Prepare a cleaning solution with water and mild soap or vinegar. Test a small area first to ensure it does not damage the leather. Gently clean the affected areas with a soft cloth or sponge.

- **Dry Thoroughly:** After cleaning, ensure the hide is thoroughly dried in a well-ventilated area to prevent mold from returning.

iii. Professional Treatment

- **Seek Expertise:** For severe mold infestations or significant damage, consider consulting a professional leather conservator or restorer. They can provide specialized treatments and repairs.

4. Preventing Physical Damage

i. Avoid Abrasive Materials

- **Handle Gently:** Handle hides with care to avoid scratches, tears, or other physical damage. Use clean, soft cloths and gloves when necessary.

- **Protective Covers:** Use protective covers or padding when storing hides to prevent contact with sharp objects or surfaces that could cause damage.

ii. Minimize Exposure to Chemicals

- **Avoid Harsh Chemicals:** Keep hides away from harsh chemicals or cleaning agents that could damage the leather. Use products specifically designed for leather care.

- **Proper Cleaning:** Clean hides using recommended methods and products to avoid unnecessary damage.

Preventing mold and damage in hides involves careful attention to drying methods, storage conditions, and handling practices. By employing proper drying techniques, maintaining a clean and controlled storage environment, and taking proactive measures to prevent and treat mold, you can ensure the quality and longevity of the leather. Regular inspections and gentle handling further protect the hide from physical damage, contributing to the overall success of the tanning and leather production process.

8.3 Long-Term Storage Solutions for Hides

Long-term storage of hides is crucial for preserving their quality, preventing damage, and ensuring they remain suitable for future use. Proper storage techniques help maintain the leather's appearance, texture, and durability over extended periods. Here's a comprehensive guide to long-term storage solutions for hides:

1. Environmental Control

i. Temperature Management

- **Cool Conditions:** Store hides in a cool environment to prevent deterioration. Ideal temperatures range from 50°F to 70°F (10°C to 21°C). Avoid extreme heat or cold, which can cause damage or affect the leather's properties.

- **Stable Temperature:** Maintain a consistent temperature to prevent fluctuations that could cause the hide to expand or contract, leading to damage.

ii. Humidity Control

- **Low Humidity:** Keep the storage area at a low humidity level, ideally between 40% and 60%. High humidity can promote mold growth and mildew, while very low humidity can lead to brittleness and cracking.
- **Dehumidifiers:** Use dehumidifiers to manage humidity levels if the storage environment tends to be damp.

iii. Ventilation

- **Air Circulation:** Ensure good ventilation in the storage area to prevent the buildup of stagnant air, which can contribute to mold growth and odor. Use fans or air circulation systems if necessary.
- **Avoiding Dampness:** Keep the area dry and well-ventilated to prevent moisture accumulation.

2. Proper Packaging

i. Breathable Materials

- **Natural Fabrics:** Use breathable packaging materials such as cotton or canvas to wrap or cover hides. These materials allow air circulation and prevent moisture buildup.
- **Avoid Plastic:** Do not use plastic sheeting or bags for long-term storage, as they can trap moisture and lead to mold and mildew growth.

ii. Protective Covers

- **Use Covers:** Protect hides with dust covers or protective wraps to shield them from dirt, dust, and physical damage. Ensure the covers are breathable to allow air flow.

- **Padding:** Use soft padding if necessary to prevent contact with hard surfaces or edges that could cause abrasions.

3. Storage Conditions

i. Storage Location

- **Cool, Dry Space:** Store hides in a cool, dry location away from direct sunlight, heat sources, and sources of moisture. Avoid basements or attics where temperature and humidity can fluctuate.

- **Avoid Direct Contact:** Keep hides off the floor and away from walls to prevent potential contact with damp surfaces or pests.

ii. Hanging vs. Flat Storage

- **Hanging:** For large hides or when space allows, hanging them on padded hangers or rods can prevent creases and folds. Ensure the hide is not overcrowded to allow for air circulation.

- **Flat Storage:** For smaller hides or limited space, store hides flat in clean, dry conditions. Place them on shelves or in drawers lined with breathable materials.

4. Regular Maintenance

i. Periodic Inspection

- **Check for Issues:** Regularly inspect stored hides for signs of damage, mold, or deterioration. Address any issues promptly to prevent further damage.
- **Reconditioning:** Occasionally recondition the leather with appropriate leather care products to maintain its flexibility and appearance.

ii. Cleaning and Treatment

- **Clean as Needed:** Clean hides according to their specific needs and care instructions. Use mild, leather-friendly cleaning products to avoid damage.
- **Conditioning:** Apply leather conditioners or treatments as recommended to keep the hide supple and prevent drying or cracking.

5. Pest Control

i. Preventing Infestations

- **Pest-Free Environment:** Ensure the storage area is free from pests such as insects or rodents that could damage the hides. Use pest control measures if necessary.
- **Regular Cleaning:** Keep the storage area clean and free from food scraps or other materials that might attract pests.

ii. Treatment for Infestations

- **Professional Help:** If an infestation is detected, seek professional pest control services to address the issue without causing further damage to the hides.

- **Preventive Measures:** Use pest repellents or treatments designed for leather storage areas to prevent future infestations.

6. Documentation and Inventory

i. Record Keeping

- **Track Storage:** Maintain records of the hides stored, including details such as type, condition, and storage location. This helps in managing and retrieving hides efficiently.

- **Inspection Logs:** Keep logs of periodic inspections and maintenance activities to track the condition of the hides over time.

ii. Labeling

- **Identify Hides:** Label storage containers or areas with relevant information about the hides, such as type, date of storage, and any special care instructions.

- **Easy Access:** Ensure labels are clear and visible to facilitate easy identification and retrieval.

Long-term storage of hides requires careful management of environmental conditions, proper packaging, and regular maintenance to preserve their quality and usability. By maintaining stable temperature and humidity levels, using breathable materials for packaging, and conducting periodic inspections, you can effectively protect hides from mold, damage, and deterioration. Implementing pest control measures and keeping accurate records further enhances the success of long-term storage, ensuring that hides remain in excellent condition for future use.

CHAPTER NINE

Advanced Techniques and Projects

9.1 Creating Leather from Tanned Hides

Creating leather from tanned hides involves several key steps that transform the tanned hide into usable leather products. This process encompasses finalizing the leather's texture, appearance, and functionality. Here's a comprehensive guide to the process of creating leather from tanned hides:

1. Preparation of Tanned Hides

i. Inspection and Sorting

- **Visual Inspection:** Examine tanned hides for defects such as uneven tanning, discoloration, or damage. Sorting hides by quality helps ensure consistent results in the final leather product.

- **Sorting:** Organize hides based on their intended use, such as upholstery, fashion, or accessories. This sorting helps in choosing the appropriate finishing techniques for each type of leather.

ii. Cleaning

- **Surface Cleaning:** Clean the tanned hide to remove any residual tanning agents, dust, or debris. Use a damp cloth or sponge to gently wipe the surface, avoiding excessive moisture that could damage the leather.

- **Conditioning:** Apply a leather conditioner to restore flexibility and suppleness, especially if the hide has been stored for a long time.

2. Leather Finishing Processes

i. Trimming and Shaping

- **Edge Trimming:** Trim any irregular edges or excess material from the hide to prepare it for further processing. Use a sharp blade or rotary cutter for clean, precise cuts.

- **Shaping:** Depending on the intended use, cut and shape the hide into desired sizes and forms. This can include cutting patterns for garments, upholstery, or accessories.

ii. Dyeing and Coloring

- **Dye Selection:** Choose dyes appropriate for the type of leather and desired color. For example, aniline dyes provide a natural look, while pigmented dyes offer more coverage and colorfastness.

- **Application:** Apply dye using methods such as immersion, spraying, or hand rubbing. Ensure even application to achieve consistent color throughout the hide.

- **Drying:** Allow the dyed hide to dry thoroughly in a well-ventilated area. Follow specific drying times recommended for the type of dye used.

iii. Surface Finishing

- **Buffing:** Use a buffing machine or soft cloth to polish the surface of the leather, enhancing its appearance and texture. Buffing can help remove excess dye and improve the leather's sheen.

- **Glossing and Sealing:** Apply a leather gloss or sealant to protect the surface and enhance the shine. This step is crucial for ensuring durability and resistance to moisture and stains.

3. Leather Softening and Conditioning

i. Softening Techniques

- **Mechanical Softening:** Use mechanical methods such as rolling or tumbling to soften the leather. This process helps to relax the fibers and improve the leather's flexibility.

- **Conditioning:** Apply a leather conditioner or softening agent to further enhance the leather's suppleness. Choose products formulated for the specific type of leather to avoid adverse reactions.

ii. Finishing Treatments

- **Oiling:** For added flexibility and protection, apply a light layer of leather oil. This helps to nourish the leather and prevent drying or cracking.

- **Protective Coatings:** Consider applying protective coatings or water repellents to increase the leather's resistance to moisture and stains.

4. Quality Control and Inspection

i. Final Inspection

- **Visual Check:** Conduct a thorough inspection of the finished leather for any defects or inconsistencies in color, texture, or surface quality. Address any issues before proceeding to the next stage.

- **Texture and Flexibility:** Test the leather for texture and flexibility to ensure it meets the desired specifications for its intended use.

ii. Testing

- **Durability Testing:** Perform tests to assess the leather's durability, including abrasion resistance and tensile strength. Ensure that the leather meets the necessary standards for its application.

- **Finish Quality:** Evaluate the finish quality, including gloss levels, smoothness, and resistance to staining or moisture.

5. Creating Leather Products

i. Pattern Cutting and Sewing

- **Pattern Preparation:** Prepare patterns or templates for the leather product you intend to create. This can include garments, accessories, or upholstery.

- **Cutting:** Use precise cutting tools to cut the leather according to the prepared patterns. Ensure accurate cutting to minimize waste and ensure a good fit.

- **Sewing:** Assemble and sew the cut pieces using appropriate thread and stitching techniques. Pay attention to seam allowances and stitching quality to ensure durability.

i. Final Assembly

- **Fitting and Assembly:** Assemble the leather pieces to create the final product. This may involve additional steps such as attaching hardware, linings, or reinforcing certain areas.

- **Finishing Touches:** Add final touches such as decorative elements, closures, or customizations to complete the product. Ensure that all components are securely attached and functional.

6. Care and Maintenance

i. Regular Care

- **Cleaning:** Clean leather products regularly using appropriate leather cleaners. Follow care instructions for the specific type of leather to avoid damage.

- **Conditioning:** Periodically apply leather conditioner to maintain suppleness and prevent dryness.

ii. Storage

- **Proper Storage:** Store leather products in a cool, dry place away from direct sunlight and heat sources. Use breathable covers or bags to protect the leather from dust and moisture.

Creating leather from tanned hides involves several essential steps: preparation, finishing, softening, conditioning, quality control, and product creation. Each step is crucial for achieving high-quality leather that meets the desired specifications for various applications. By following these processes carefully and maintaining proper care and maintenance, you can ensure that the final leather products are durable, attractive, and functional.

9.2 Crafting with Fur and Leather

Crafting with fur and leather offers a wide range of creative possibilities, from functional items like clothing and accessories to decorative pieces. Both materials require specific handling and techniques to ensure quality and durability. Here's a comprehensive guide to crafting with fur and leather, including tips, techniques, and project ideas.

1. Understanding Fur and Leather

i. Characteristics of Fur

- **Texture and Appearance:** Fur has a soft, plush texture with a distinctive grain pattern. The pile or hair length varies depending on the animal.

- **Care Requirements:** Fur requires careful handling and maintenance to avoid matting, shedding, or damage.

ii. Characteristics of Leather

- **Durability:** Leather is a strong, flexible material with various finishes and textures. It can be full-grain, top-grain, or split, each offering different qualities.
- **Tanning Process:** The type of tanning affects the leather's softness, durability, and appearance.

2. Preparation and Planning

i. Choosing Materials

- **Fur Selection:** Choose high-quality fur based on your project's requirements. Consider the type of fur, length, color, and condition.
- **Leather Selection:** Select leather that suits the intended use, such as soft, supple leather for garments or thicker, more durable leather for accessories.

ii. Tools and Supplies

- **For Fur:** Sharp scissors, seam ripper, fur glue, needle and thread, and furrier's tape.
- **For Leather:** Rotary cutter or sharp utility knife, cutting mat, leather stitching tools, mallet, and leather adhesive.

iii. Pattern Making

- **Create or Obtain Patterns:** Develop or acquire patterns suitable for your project. Patterns should be accurate and tailored to the specific dimensions and shapes of fur and leather.

- **Adjust for Materials:** Consider the thickness and grain direction of the materials when adapting patterns. Make necessary adjustments to ensure a proper fit and finish.

3. Cutting and Shaping

i. Cutting Fur

- **Use Sharp Scissors:** Use sharp, high-quality scissors designed for cutting fur to achieve clean edges. Avoid using regular household scissors as they can damage the fur.

- **Cutting Direction:** Cut in the direction of the fur's grain to prevent uneven edges or damage. Avoid cutting through the fur directly; instead, use a slight angle to maintain a smooth finish.

ii. Cutting Leather

- **Use a Rotary Cutter or Knife:** A rotary cutter or sharp utility knife provides clean, precise cuts. Ensure the cutting surface is suitable and stable.

- **Cutting Techniques:** Use a straight edge or ruler for straight cuts. For curves, use a flexible curve ruler or carefully maneuver the knife to achieve the desired shape.

4. Assembling and Stitching

i. Joining Fur and Leather

- **Sewing Fur:** Use a heavy-duty needle and thread or a sewing machine designed for fur. Sew slowly to prevent puckering or shifting of the fur pile.

- **Joining Leather:** Leather can be stitched by hand or machine. Use a strong thread and appropriate needle for leatherwork. For heavier leather, use a leather sewing machine or hand-stitching tools.

ii. Adhering Materials

- **Fur Adhesives:** Use specialized fur glue or adhesive designed for fur to join fur pieces. Apply adhesive sparingly to avoid excess glue on the fur.

- **Leather Adhesives:** Use leather-specific adhesives or contact cement to bond leather pieces. Apply a thin, even layer of glue and press the pieces together firmly.

5. Finishing Touches

i. Edges and Seams

- **Fur Edges:** Trim fur edges carefully to ensure a neat finish. Consider using fur binding or hem tape to prevent fraying.

- **Leather Edges:** Burnish leather edges using an edge slicker or bone folder to create a smooth, rounded edge. Apply edge dressing or sealer to protect the edges.

ii. Adding Hardware

- **For Fur Projects:** Attach hardware such as snaps or buttons carefully to avoid damaging the fur. Use small, sharp tools to create necessary holes.

- **For Leather Projects:** Install hardware like rivets, grommets, or buckles using appropriate tools and techniques. Ensure that all hardware is securely attached and functional.

iii. Cleaning and Maintenance

- **Fur Care:** Brush fur regularly to maintain its appearance and prevent matting. Store fur items in a cool, dry place away from direct sunlight.

- **Leather Care:** Clean leather with a damp cloth and use leather conditioners to maintain suppleness. Store leather items in a cool, dry area and avoid excessive moisture or heat.

6. Project Ideas

i. Clothing and Accessories

- **Fur Coats and Jackets:** Create stylish and warm outerwear by combining fur with leather accents. Pay attention to pattern alignment and fur placement.

- **Leather Bags and Wallets:** Craft durable and functional leather bags, wallets, or purses. Use high-quality leather and reinforce stress points with additional stitching or rivets.

ii. Home Décor

- **Fur Pillows and Throws:** Add luxury to home décor with fur pillows, throws, or seat covers. Choose fur that complements existing furnishings and décor styles.

- **Leather Furniture Covers:** Create custom leather covers or slipcovers for furniture. Ensure a snug fit and use appropriate fastenings or adhesives.

iii. Craft Projects

- **Fur Trimmings:** Use fur trimmings to add decorative accents to clothing, accessories, or home décor items. Consider combining fur with other materials for a unique look.

- **Leather Keychains and Coasters:** Craft small leather items like keychains or coasters as gifts or personal accessories. Experiment with different leather finishes and designs.

Crafting with fur and leather involves careful selection of materials, precise cutting and shaping, and skilled assembly techniques. By understanding the unique properties of each material and employing appropriate tools and methods, you can create a wide range of high-quality, functional, and decorative items. From clothing and accessories to home décor and craft projects, the possibilities are vast and versatile, allowing for creativity and personalization in every piece.

9.3 Ethical Trapping and Hunting Community Involvement

Community involvement in ethical trapping and hunting is essential for promoting responsible practices, educating others, and fostering a culture of conservation and respect for wildlife. Engaging with local, national, and global communities helps ensure that trapping and hunting are conducted in ways that are sustainable, humane, and aligned with conservation goals. Here's a comprehensive guide to community involvement in ethical trapping and hunting:

1. Education and Awareness

i. Educational Programs and Workshops

- **Local Workshops:** Organize or participate in local workshops and seminars to educate the community about ethical trapping and hunting practices. Topics may include humane trapping techniques, wildlife conservation, and the importance of adhering to regulations.
- **School Programs:** Collaborate with schools to integrate wildlife education into curricula. Provide age-appropriate materials and activities that promote understanding and respect for wildlife.

ii. Public Outreach

- **Community Events:** Participate in or host community events such as wildlife fairs, conservation days, or outdoor expos. Use these events to share information about ethical trapping and hunting, and to showcase the benefits of responsible practices.

- **Media Campaigns:** Utilize local media, social media platforms, and community newsletters to spread awareness about ethical trapping and hunting. Share success stories, educational content, and information about local regulations.

iii. Collaboration with Organizations

- **Partnerships:** Partner with wildlife conservation organizations, hunting clubs, and environmental groups to promote ethical practices. Joint efforts can amplify the message and reach a wider audience.

- **Resource Sharing:** Share resources, best practices, and educational materials with organizations that have similar goals. Collaboration can enhance the effectiveness of outreach and education efforts.

2. Advocacy and Policy Influence

i. Engaging with Policy Makers

- **Legislative Advocacy:** Work with policymakers to advocate for laws and regulations that support ethical trapping and hunting practices. Provide evidence-based recommendations and participate in public consultations or hearings.

- **Policy Education:** Educate legislators and regulators about the benefits of ethical practices and the need for well-informed policies. Offer to provide expert testimony or serve on advisory committees.

ii. Community Input and Feedback

- **Surveys and Forums:** Conduct surveys or community forums to gather input from local residents about trapping and hunting practices. Use this feedback to inform advocacy efforts and to ensure that community concerns are addressed.

- **Public Consultation:** Encourage open dialogues with the community about ethical trapping and hunting. Address concerns, answer questions, and provide information to build trust and support.

iii. Promoting Conservation Initiatives

- **Support Conservation Projects:** Get involved in or support local conservation projects that align with ethical trapping and hunting principles. This could include habitat restoration, wildlife monitoring, or educational outreach.

- **Funding and Donations:** Contribute to or fundraise for conservation organizations that work to promote ethical practices and protect wildlife habitats.

3. Best Practices and Ethical Standards

i. Adherence to Regulations

- **Compliance:** Ensure that all trapping and hunting activities adhere to local, state, and national regulations. Stay informed about changes in laws and adjust practices accordingly.

- **Reporting Violations:** Encourage community members to report violations or unethical practices. Establish channels for reporting and ensure that violations are addressed promptly.

ii. Ethical Training and Certification

- **Certification Programs:** Promote or participate in certification programs that focus on ethical trapping and hunting practices. These programs can provide formal recognition of responsible practices and enhance credibility.

- **Training Workshops:** Offer or attend training workshops that cover topics such as humane trapping techniques, animal handling, and conservation principles.

iii. Promoting Humane Practices

- **Humane Trapping:** Advocate for and practice humane trapping methods that minimize animal suffering. Use traps that are designed to be effective yet gentle, and regularly check traps to ensure prompt removal of captured animals.

- **Ethical Hunting:** Emphasize the importance of ethical hunting practices, including respect for the animal, proper shot placement, and minimizing waste. Promote the use of all parts of the animal to honor the life taken.

4. Community Involvement and Volunteer Opportunities

i. Volunteer Programs

- **Local Volunteering:** Encourage community members to volunteer for organizations or projects related to ethical trapping and hunting. Volunteering opportunities may include habitat restoration, educational outreach, or wildlife monitoring.

- **Mentorship:** Offer mentorship programs for new trappers or hunters. Provide guidance on ethical practices, safety, and conservation principles to foster responsible behavior from the outset.

ii. Community Groups and Clubs

- **Join or Form Groups:** Join existing community groups or clubs focused on trapping and hunting. Alternatively, form new groups to create a network of individuals committed to ethical practices.

- **Group Activities:** Organize group activities such as clean-up events, conservation projects, or educational workshops. Foster a sense of community and shared responsibility for ethical practices.

iii. Recognition and Incentives

- **Recognize Efforts:** Acknowledge and celebrate individuals or groups that demonstrate exemplary ethical practices. Recognition can include awards, public acknowledgment, or certificates of appreciation.
- **Incentives:** Provide incentives for community members to engage in ethical trapping and hunting practices. This could include discounts on gear, access to exclusive events, or other rewards.

5. Fostering Respect and Collaboration

i. Building Relationships

- **Respectful Dialogues:** Foster respectful dialogues with all stakeholders, including hunters, trappers, conservationists, and the general public. Build mutual understanding and collaboration to achieve common goals.
- **Cultural Sensitivity:** Recognize and respect cultural practices related to trapping and hunting. Engage with diverse communities to understand their perspectives and incorporate their values into ethical practices.

ii. Collaborative Projects

- **Joint Initiatives:** Collaborate on joint initiatives that promote ethical trapping and hunting. This could include research projects, conservation efforts, or community outreach programs.

- **Shared Goals:** Work together with other organizations and community groups to achieve shared goals related to wildlife conservation and ethical practices.

Community involvement in ethical trapping and hunting is crucial for promoting responsible practices, influencing policy, and fostering a culture of conservation and respect for wildlife. By engaging in education, advocacy, and collaborative efforts, individuals and groups can contribute to the preservation of ethical standards and support sustainable practices. Active participation in community initiatives, adherence to regulations, and promotion of humane practices help ensure that trapping and hunting are conducted in a manner that benefits both wildlife and communities.

Thanks For Reading!

www.ingramcontent.com/pod-product-compliance
Lightning Source LLC
Chambersburg PA
CBHW071921210526
45479CB00002B/511